AA003343

Process Development Symposium 2013

Process Scale-Up and its Challenges

Oak Brook, Illinois, USA
11 – 13 June 2013

ISBN: 978-1-5108-1604-6

Printed from e-media with permission by:

Curran Associates, Inc.
57 Morehouse Lane
Red Hook, NY 12571

Some format issues inherent in the e-media version may also appear in this print version.

Copyright© (2013) by AIChE
All rights reserved.

Printed by Curran Associates, Inc. (2015)

For permission requests, please contact AIChE
at the address below.

AIChE
120 Wall Street, FL 23
New York, NY 10005-4020

Phone: (800) 242-4363
Fax: (203) 775-5177

www.aiche.org

Additional copies of this publication are available from:

Curran Associates, Inc.
57 Morehouse Lane
Red Hook, NY 12571 USA
Phone: 845-758-0400
Fax: 845-758-2634
Email: curran@proceedings.com
Web: www.proceedings.com

TABLE OF CONTENTS

OPENING KEYNOTE

PROCESS DESIGN, MODELING, AND FLOWSHEETS

PREVALENT INDUSTRY UNIT OPERATIONS

SEPARATIONS UNIT OPERATIONS YOU DIDN'T LEARN ABOUT IN SCHOOL

PILOT PLANTS

ENVIRONMENTAL, HEALTH, AND SAFETY CONCERNS

CASE STUDIES

PROCESS SCALE-UP DOS & DON'TS

Innovative Use of Industrial Waste Gases to Produce Sustainable Fuels & Chemicals

LanzaTech

The LanzaTech Process

LanzaTech

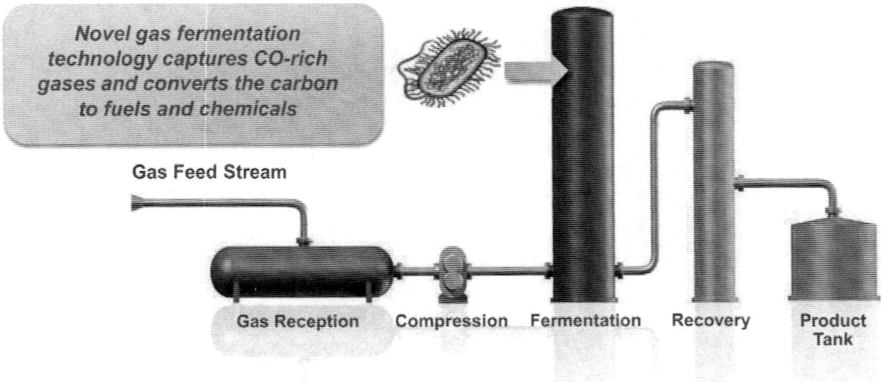

- **Gases are sole source of energy**
- **Production of fuels and chemicals**
- **Potential to make <u>material</u> impact on the future energy pool (>100s of billions of gallons per year)**

No impact on water, food, land or biodiversity

2

Energy Demand Rising

LanzaTech

Population growth and climate change put pressure on land, food, water, and ecosystems…

Share of Global Energy Demand

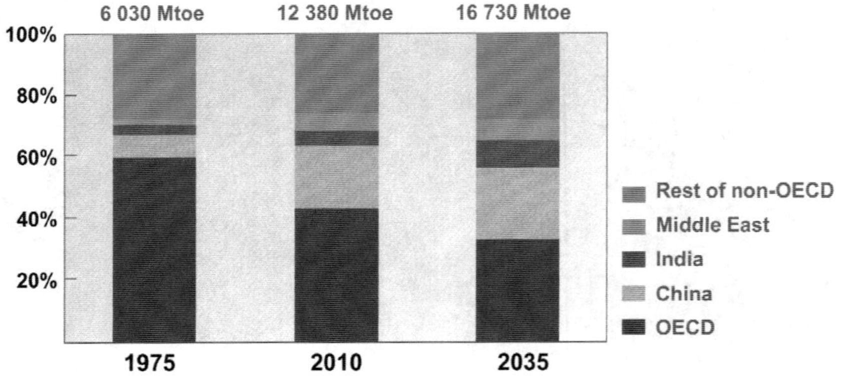

IEA World Energy Outlook 2012

3

Carbon Emissions Rising

LanzaTech

Containing CO_2 growth to safe atmospheric levels (below 450 ppm) requires that zero carbon fuels make up > 30% of the fuel pool. Current Rate: ~8 billion tonnes of CO_2 per year

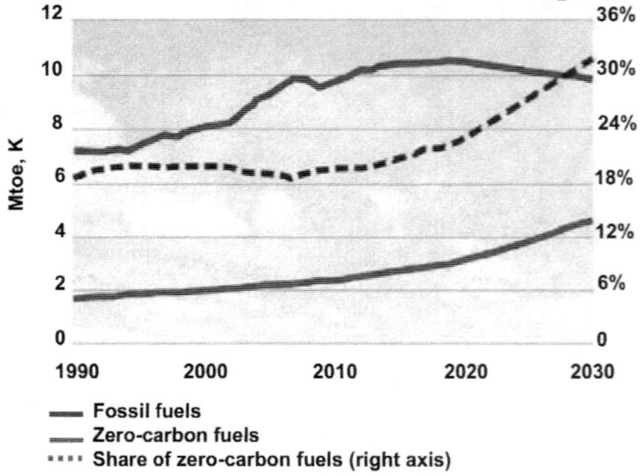

Source: IEA world Energy Outlook 2011

Data from NOAA* taken at Mauna Loa, Hawaii for May 10 2013 shows CO_2 levels have exceeded 400 ppm

* National Oceanic and Atmospheric Administration

4

Despite Large Energy Access Gap

LanzaTech

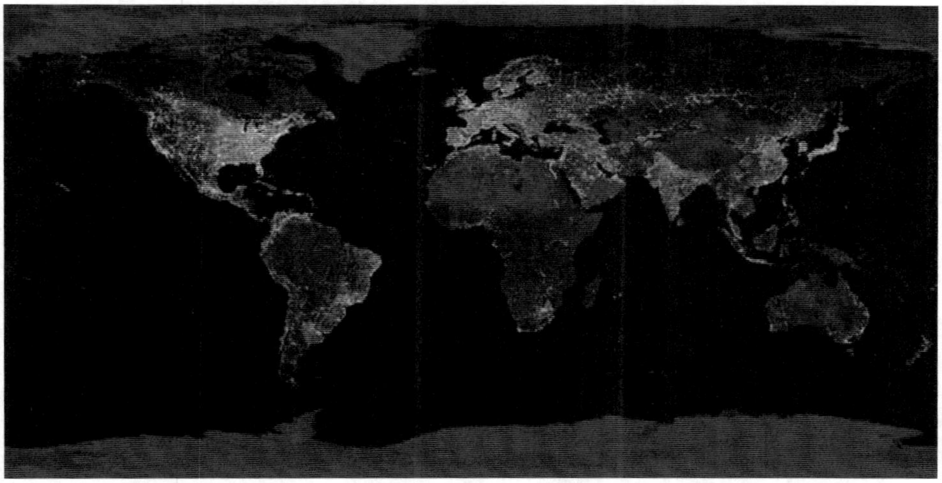

1.3B people still lack access to basic energy

5

Must Diversify our Energy Basket

LanzaTech

But Solution Must Be Sustainable to Have a Seat at the Table

6

Must Promote Resource Efficiency

LanzaTech

Energy can be Carbon free

Wind:

Solar:

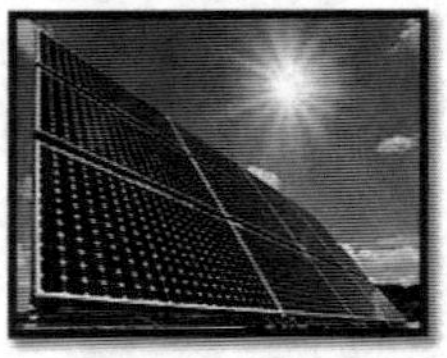

Hydro:

- **Liquid Fuels & Petrochemicals** must contain:

7

Gas Fermentation:
Ancient biology for a modern need

LanzaTech

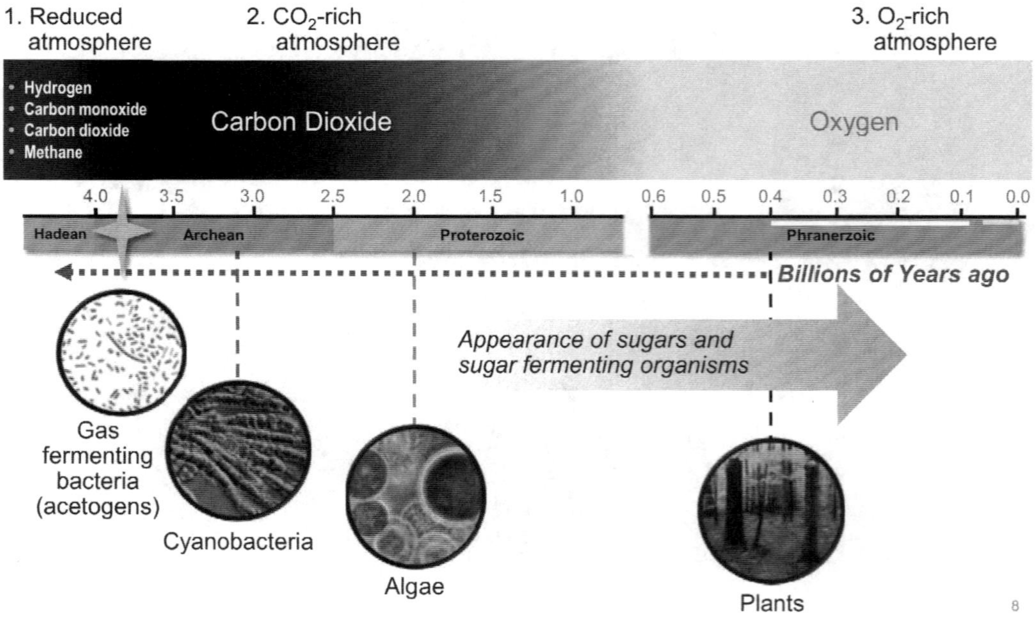

8

Leveraging the Early Organisms

LanzaTech

- The <u>biology</u> allowing reducing gases to provide *all* the carbon and energy for life still exists today in a class of bacteria called *Acetogens*.
- Billions of years ago gases from hydrothermal vents were used for growth.
- Today similar gases are wasted or emitted into the atmosphere.

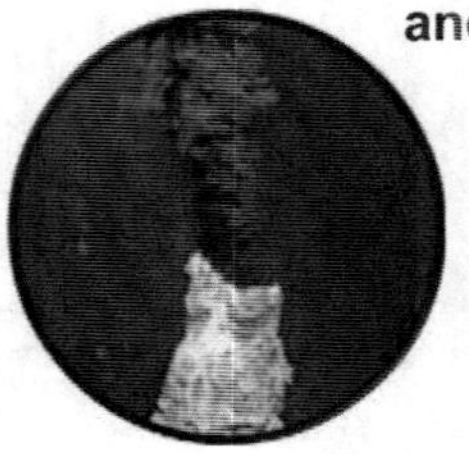

Gas from both *hydrothermal vents* and *steel manufacturing* include:

- Carbon monoxide (CO)
- Hydrogen (H_2)
- Carbon dioxide (CO_2)
- Hydrogen sulphide (H_2S)
- Methane (CH_4)

LanzaTech harnesses these ancient biological organisms to make fuels and chemicals

9

The LanzaTech Solution

LanzaTech

- LanzaTech takes advantage of this ancient biological process to make sustainable fuels and chemicals
- Our microbes use unavoidable gaseous residues as their sole energy and carbon source
- Potential to make material impact on the future energy pool (>100s of billions of gallons per year)

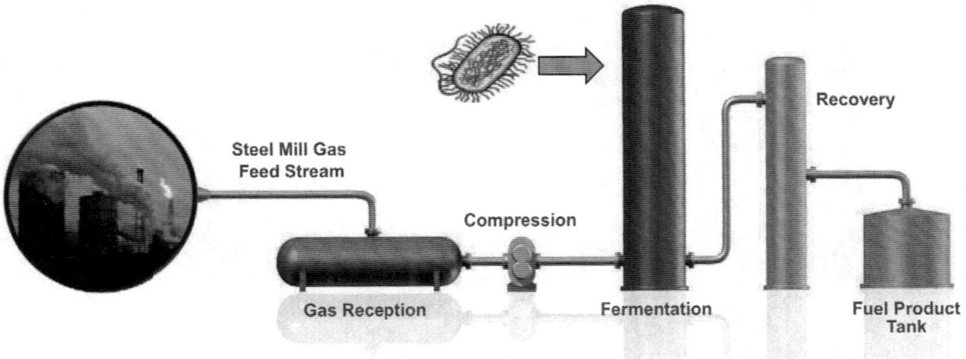

No impact on water, food, land or biodiversity

10

Fast Path to Commercial Scale

LanzaTech

LanzaTech — BAOSTEEL

- Pre-commercial facility in operation in Shanghai for >8 months meeting and exceeding all its performance targets and milestones
- Capacity 100,000 gallon/year ethanol
- Technology has been approved in China for commercial deployment, by the NDRC

LanzaTech — 首钢集团 SHOUGANG GROUP

- Operation of additional 100,000 gallon/year plant with second Chinese Partner, Shougang Group, in Beijing
- Sustainability Assessment of Beijing plant currently underway with RSB.

Co-locating LanzaTech's Technology Steel Mill brownfield sites reduces land footprint, improves economics and reduces construction time

11

Carbon Capture and Utilization

LanzaTech

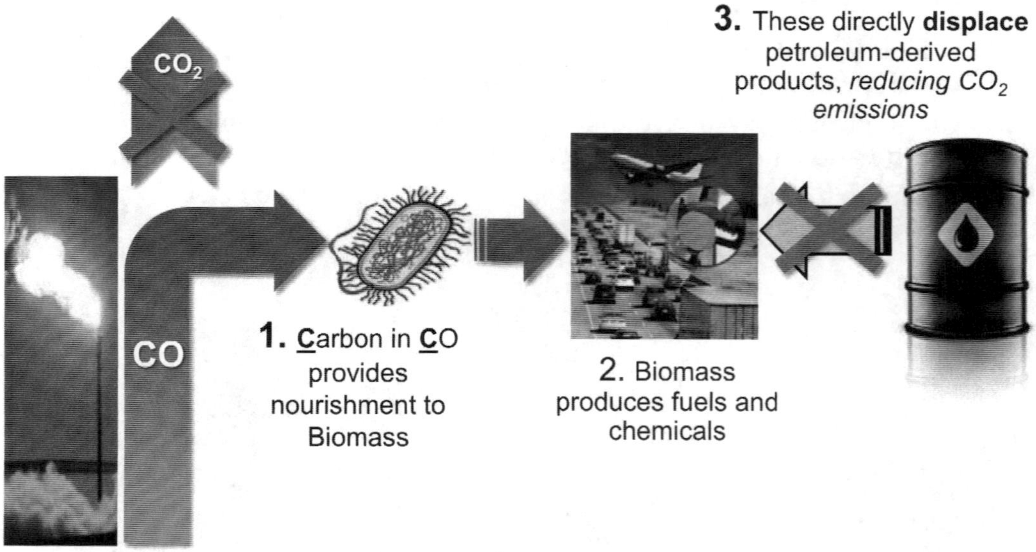

** carbon capture **before CO_2 production***
** beneficial reuse to displace petroleum-derived products*

12

By the Numbers

LanzaTech

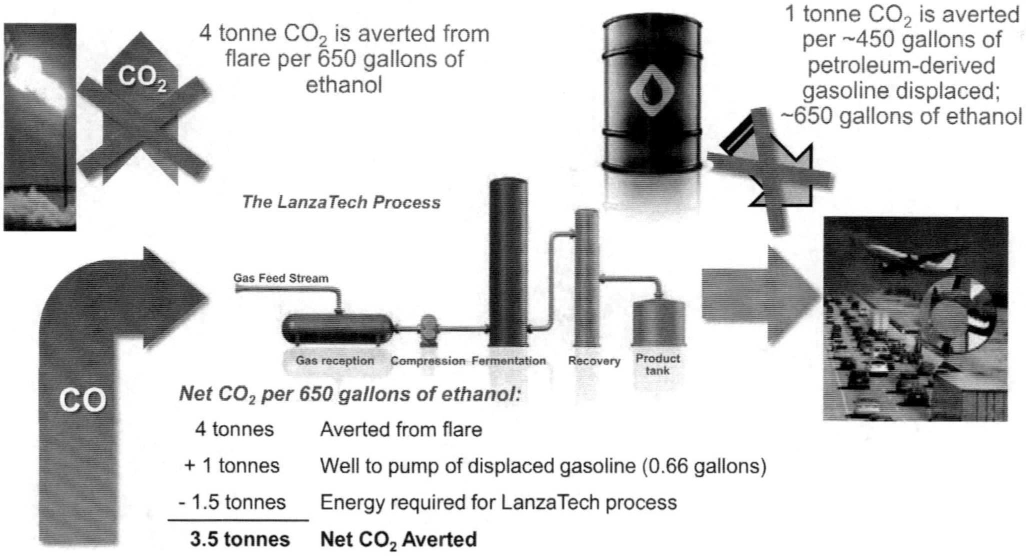

Potential from Steel Industry ~ 30 Billion Gallons of Ethanol
Impact: ~150 Million tonnes of CO_2 Averted

13

New Route to Fuels and Chemicals

LanzaTech

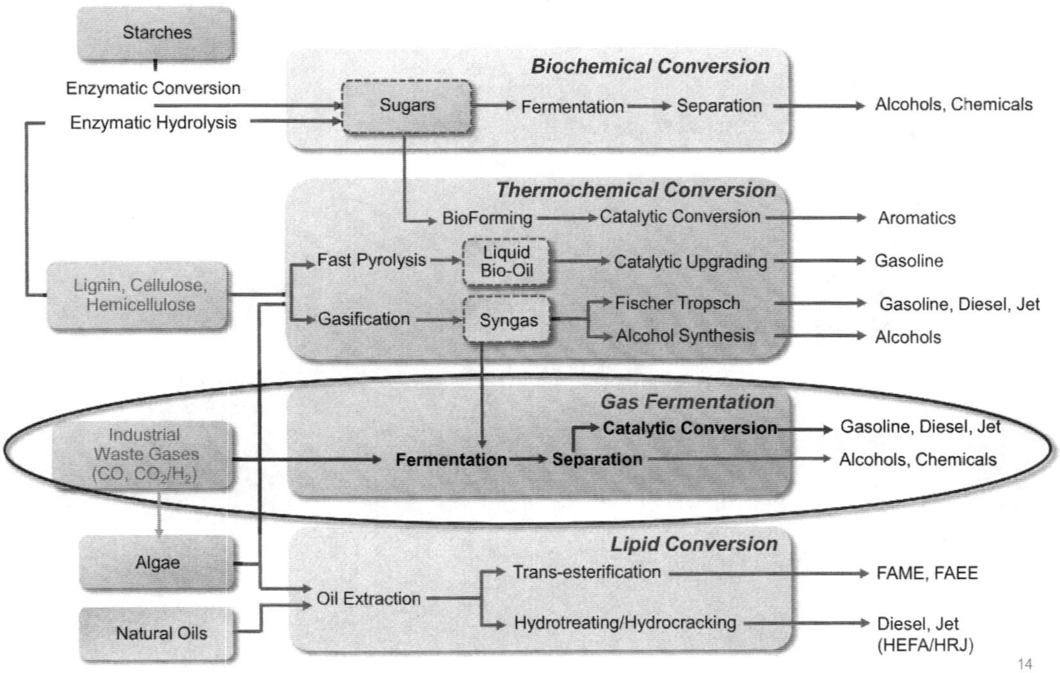

14

Accessible Feedstock Pool

LanzaTech

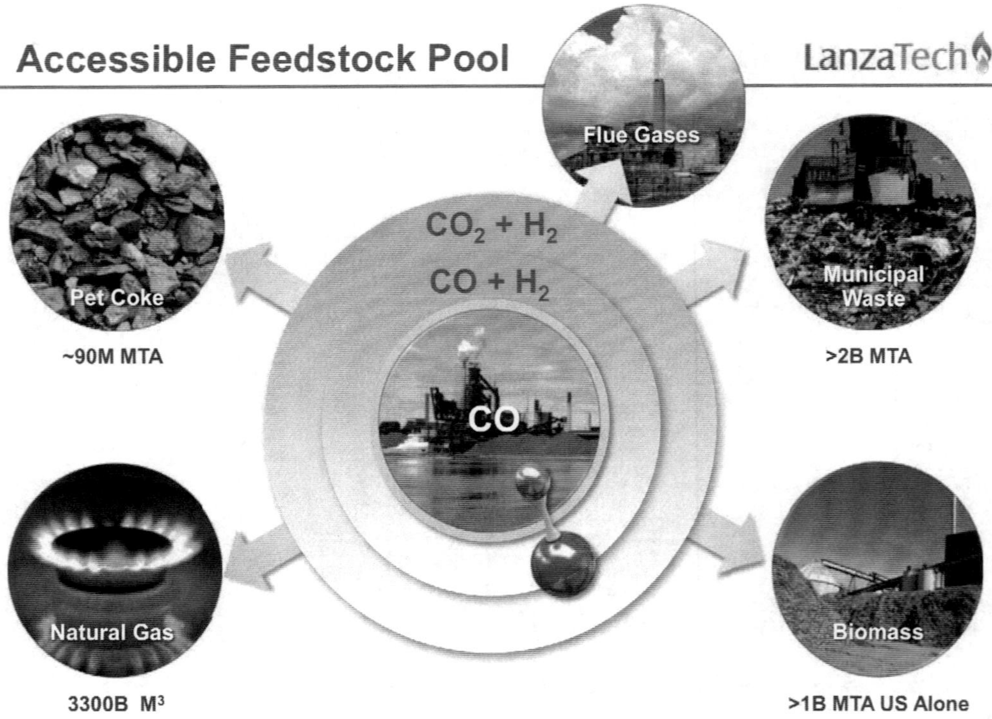

*2010 production data – IEA, UNEP

Product Portfolio

LanzaTech

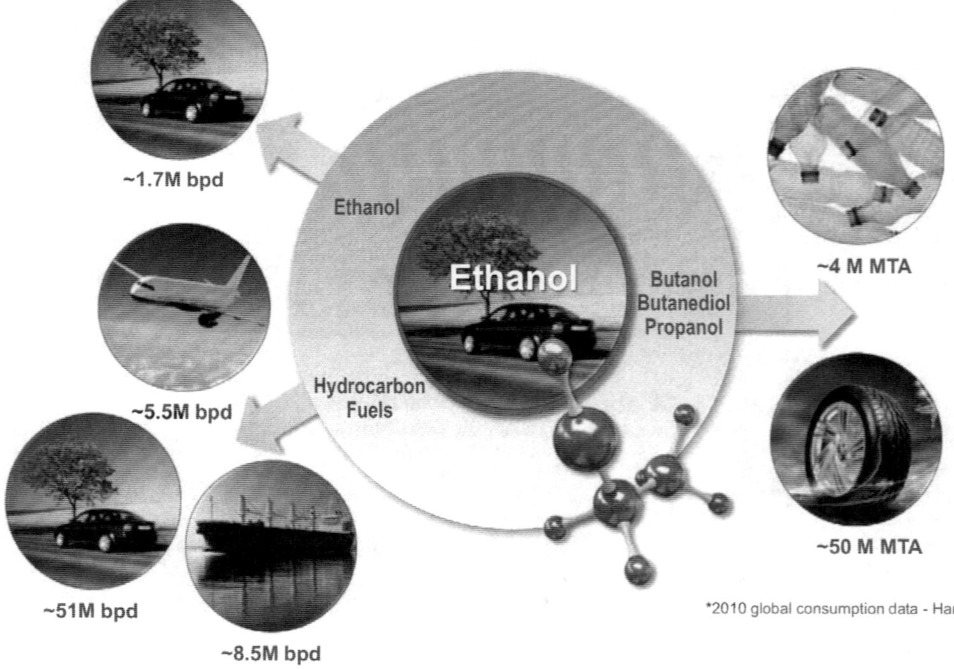

*2010 global consumption data - Harts, IEA

Commercialization of Aviation Fuel

LanzaTech

LanzaTech

BOEING

Imperial College of London

RSB ROUNDTABLE ON SUSTAINABLE BIOFUELS

Swedish BioFuels

virgin atlantic

Team Work is Key to Success

INVISTA Partnership

LanzaTech

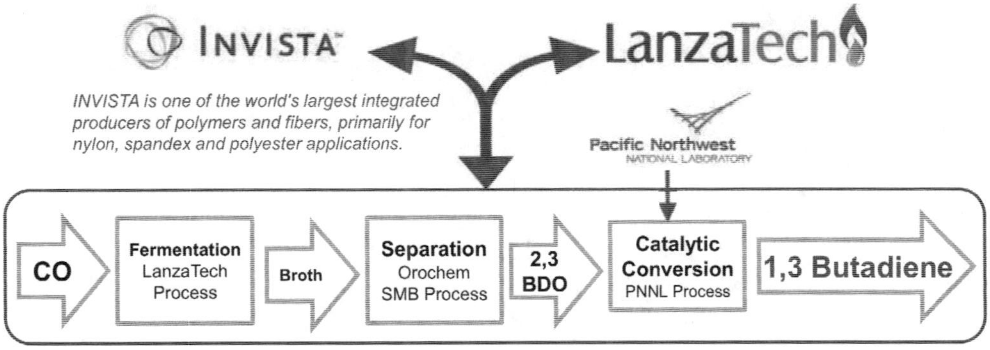

INVISTA and LanzaTech have established a strategic partnership to:

1. Develop a Two Step process to produce 1,3 Butadiene
2. Develop an alternate Direct single step route to 1,3 Butadiene

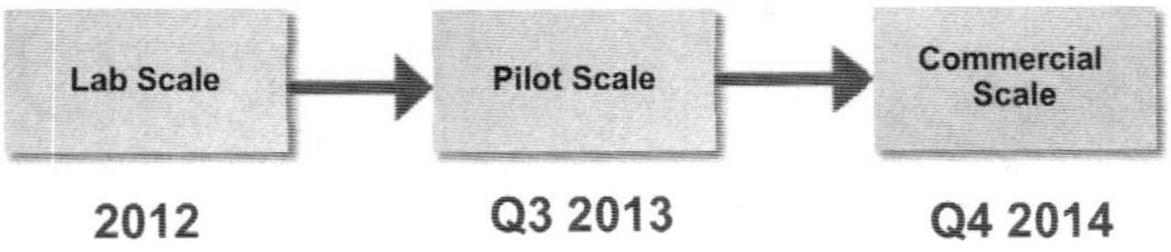

18

A Few Predictions

LanzaTech

1876	"This 'telephone' has too many shortcomings to be seriously considered as a means of communication" *Western Union Memo*
1895	"Heavier-than-air flying machines are impossible" *Lord Kelvin, President Royal Society*
1920	"The wireless music box (radio) has no imaginable commercial value" *David Sarnoffs Associates in response to his urgings for investments in the radio*
1943	"I think there's a world market for maybe five computers" *Thomas Watson, Chairman IBM*
1949	"Computer in the future may weigh no more than 1.5 tons" *Popular Mechanics forecasting the relentless march of science*
1977	"There is no reason anyone would want a computer in their home" *Ken Olson, President, Chairman and Founder of Digital Equipment*
1981	"640K ought to be enough computer memory for anyone" *Bill Gates*

19

LanzaTech

Predictions are simply extrapolations of the past…

…innovation expands the 'art of the possible'

…today's 'unimaginable' is tomorrow's 'conventional wisdom.'

20

LanzaTech

"It is very difficult to predict the future, it is much easier to invent it."

Alan Key

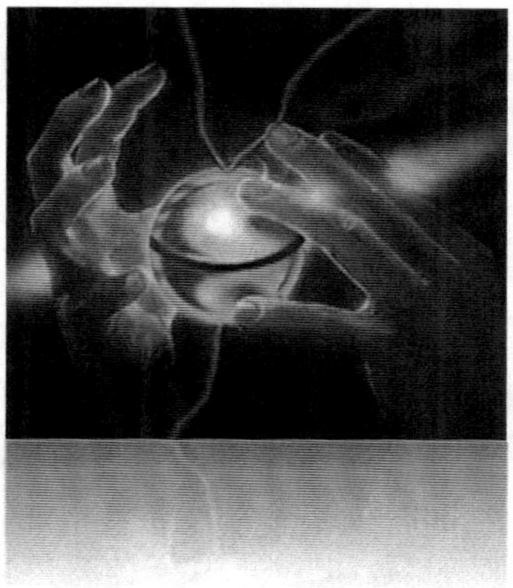

21

LanzaTech

"...relevant scale"

"...relevant cost"

"...relevant adoption"

22

Early Stage Process Development: *Accelerating Process Scale-up*

LanzaTech

Requires Science/Engineering Interaction

Decisions
- Reaction Conditions
- Solvent type/amount
- Catalyst type/amount
- Co-feed type/amount
- Separation process /conditions
- ....

→

Process Design
- Product yield/purity
- Byproduct type/amount
- Waste stream type/amount
- Equipment type/size
- Utility type/magnitude
- ...

→ ***Economics/ Sustainability***

Should be More Collaborative, Less Iterative

- **Decisions drive the process design, but the economics and sustainability assessment then inform the decision process**
- **Need fairly accurate estimates of process design, economics and impact on sustainability, but done with a minimal amount of time, effort, and money**

23

Things to Watch Out For

LanzaTech

Doing the Wrong Analysis →
- **Shutting the Project Down Too Early**
- **Letting the Project go to Far**

Doing too Much Analysis →
- **Getting caught up in details**
- **Not moving fast enough**
- **Not able to adapt**

Not Understanding the Problem, or Market, or Data →
- **Missing the opportunity**
- **Over/underestimating the opportunity**

24

Scale-up Considerations

LanzaTech

- **Technical**
  - Getting to scale quickly is critical
  - What are the limits in our scale, and how do we overcome these quickly
  - "Don't let the perfect be the enemy of the good"
- **Regulatory. Will my process/product meet local regulatory standards?**
- **Government. What government approvals are needed?**
- **LCA. Will my process meet local and global LCA standards?**

25

Scaling Up LanzaTech's Technology

LanzaTech

Lab	Pilot	Demo	Commercial
2005	2008	2012	2014
~0.5-1 kg/day	50-100 kg/day	~1500 kg/day	~300,000 kg/day

Fast Path to Commercialization

26

Technoeconomic Assessment (TEA): A Key Tool in Process Commercialization

LanzaTech

Inputs
- Gas Flow & Composition
- Regional Costs

Process Model
- Software Based (Excel and HYSYS Base)
- Incorporates Pilot and Demo Operational Results

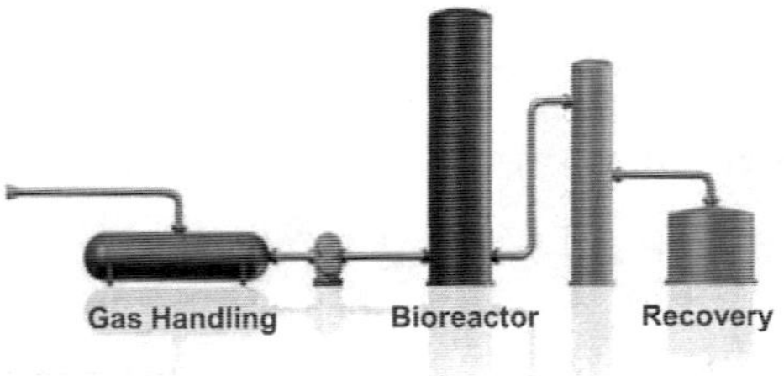

Outputs
- OPEX & CAPEX
- Economic Evaluation (IRR)
- Sensitivity Analysis
- LCA

27

Modeling to Support the Scale Up Effort

LanzaTech

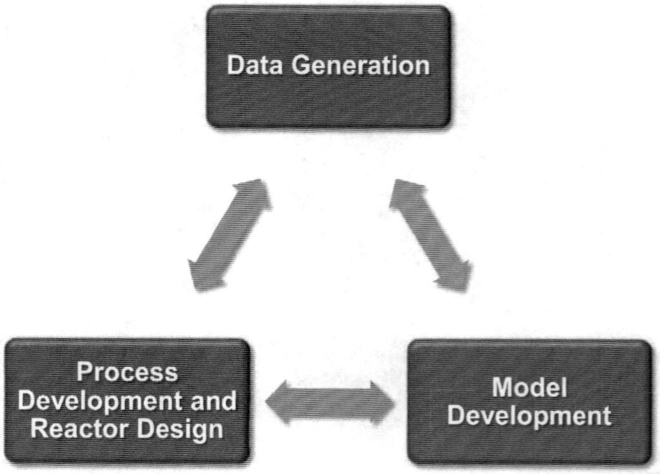

28

Model Development

LanzaTech

- **Hydrodynamic model**
  - Interaction of feed gas bubbles in liquid (1-D)
  - 3-D CFD model for detailed internals design
- **Kinetic models**
  - Input-response model
  - Structured model
  - Genomic model
- **Interaction of models**
  - Built into TEA models
  - Built into Engineering tools

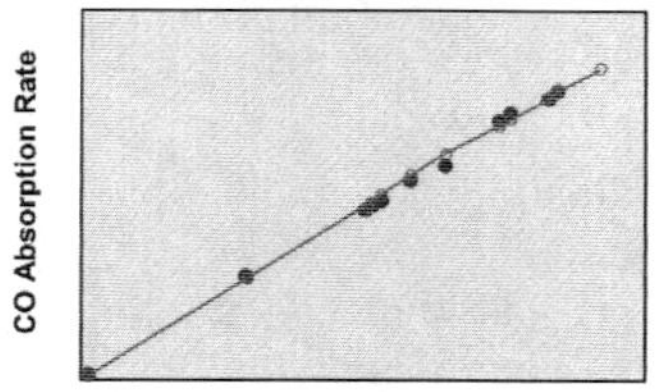

Utilize tools at different scales to generate the data required, test ideas and concepts, and generate models

29

LanzaTech Gas to Liquid Platform

LanzaTech

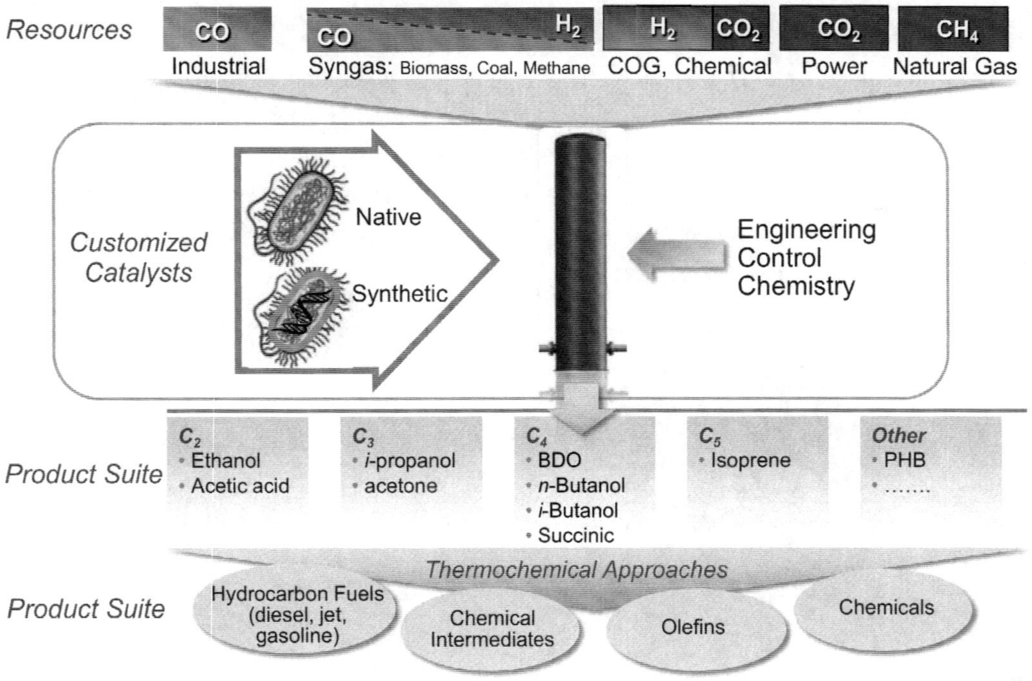

30

LanzaTech Global Partnerships

LanzaTech

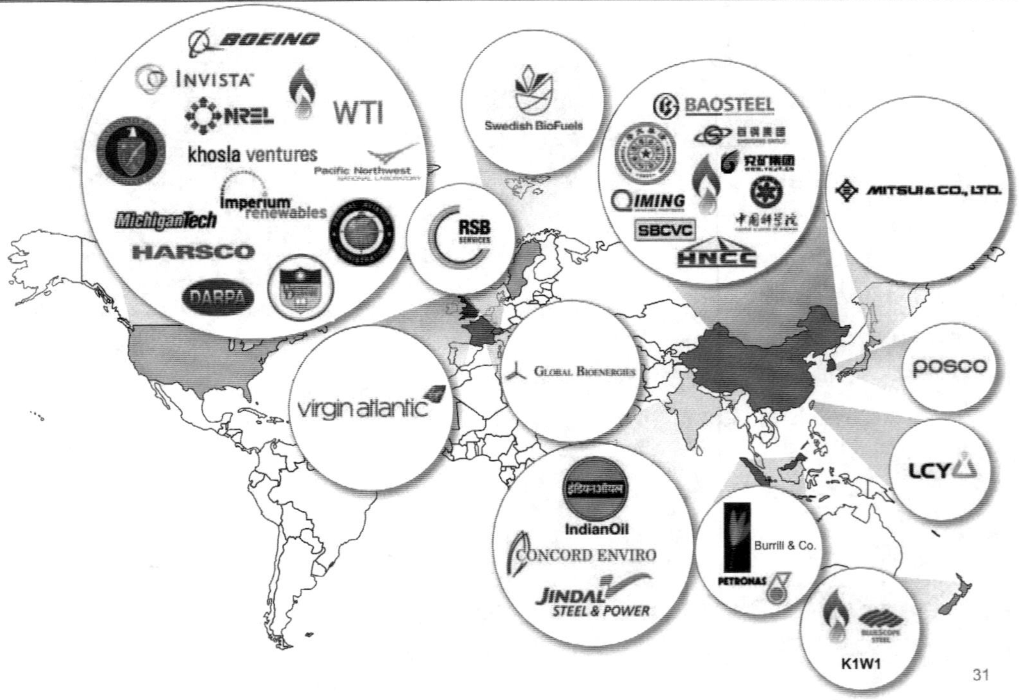

31

LanzaTech

"We choose to go to the moon. We choose to go to the moon in this decade and do the other things, not because they are easy, but because they are hard, because that goal will serve to organize and measure the best of our energies and skills, because that challenge is one that we are willing to accept, one we are unwilling to postpone, and one which we intend to win."

John Kennedy, 1962

32

SE4All

LanzaTech

SUSTAINABLE ENERGY FOR ALL is an initiative launched by the United Nations Secretary-General to make sustainable energy for all a reality by 2030.

"Achieving sustainable energy for all is not only possible, but necessary. It is the golden thread that connects development, social inclusion and environmental protection."

Secretary-General Ban Ki-moon, 2012

33

LanzaTech

the Game Changer...

34

Incorporate Transport Calculations to Increase Reality of Thermodynamic Simulations

Chemstations™

David Hill
Fred C. Justice, P.E., F.AIChE
www.chemstations.com

Premise

Thermodynamic simulations provide a critical but limited basis for process design. Adding transport calculations can produce results that agree more closely with data obtained from actual operations.

Chemstations™ 2

Introduction

- Three examples, reflecting increasing complexity and consequent reduction in frequency of use, will support this premise:
  - Simultaneous heat loss and pressure drop in a pipe
  - Coupled calculation of heat transfer simulations and heat exchanger detailed process design
  - Mass transfer-dominated vapor/liquid multistage distillation calculations

Chemstations™ 3

The core of all first-principle steady-state simulations is thermodynamic equilibrium flash.

- Equilibrium = infinite time (and equipment size)
- No composition or temperature profile in most operations
- Capability definitions are ideal:
  - Rotating equipment (pumps compressors) output independent of pressure
  - Heat exchanger U independent of flow
  - Staged operations are theoretical, or have defined efficiency
  - In always = out (no surges or imbalances)
  - Thermal definitions ideal (adiabatic isothermal)
- For these reasons, thermodynamic simulations must be considered 'first approximations'

Chemstations™ 4

Can we do better?

- Yes, add more transport calculations!
- PCs are faster now, so computational complexity is less of a problem.
- Simulations have always had some transport calculations.

Chemstations™ 5

Pipe flow example

- Heat and Momentum Transfer in a pipe system results in VLE changes

Consider two schedule 40 bare pipes in series, each 650 feet long, carrying a mixture of condensable gases just above the dew point at 255 °F and 50 psia.

The first pipe is 4” in diameter, and the second is 3”.

Chemstations™ 6

Momentum and heat transfer in pipes

Plotting conditions every 50 feet shows a more rapid pressure drop in the 3” segment, as would be expected.

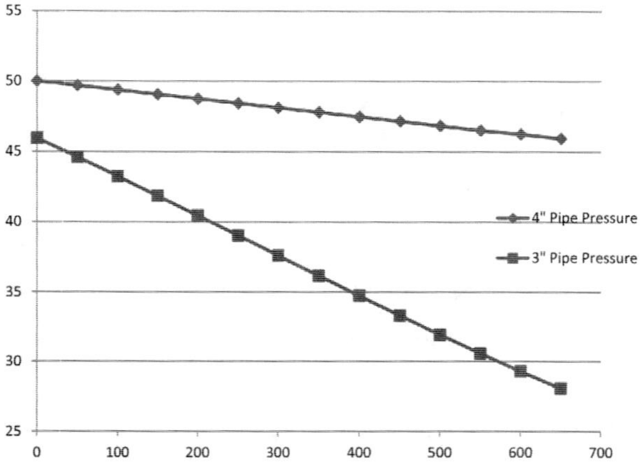

Chemstations™ 7

Momentum and heat transfer in pipes

When a similar plot of temperature every 50 feet is made, the temperature drop suddenly slows in the 4” pipe, but picks back up again in the 3” pipe.

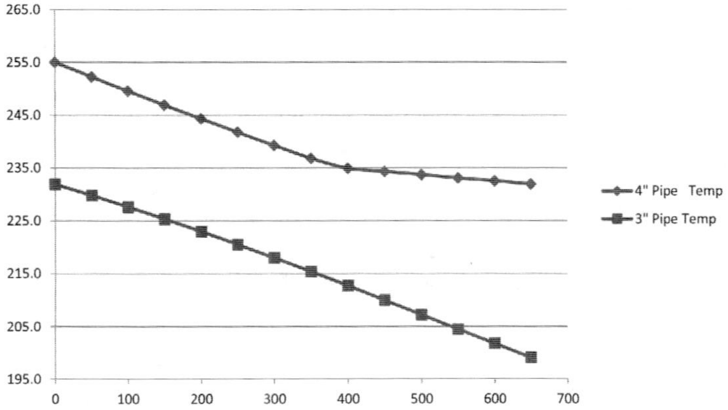

Chemstations™ 8

Momentum and heat transfer in pipes

A final plot of the vapor percentage along the length of the pipe offers some insight:

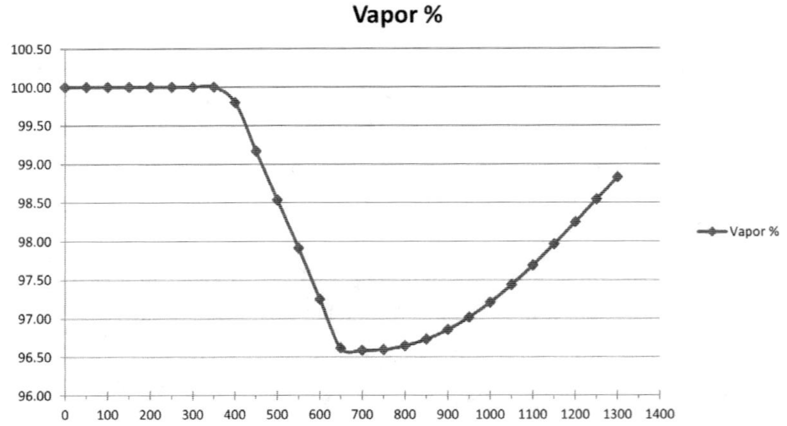

Chemstations™

9

Momentum and heat transfer in pipes

- Up to ~350 feet, the heat loss reduces the superheat of the stream, but as condensation begins, the heat lost is partially compensated by condensing vapor. This continues until 650 feet, when the higher pressure drop and smaller area for heat loss in the 3” pipe pushes the mixture back toward the dew point pressure, thereby boiling off some of the condensed phase.
- Simultaneous heat and momentum transfer can add understanding to a VLE calculation.

Chemstations™

10

Coupled thermodynamic and transport design of a heat exchanger

- Thermodynamic simulators calculate overall heat requirements, and can segment devices by heat and condensation curves.
- These curves allow transport properties to be calculated in each segment with increased accuracy.
- A process design can be established, and/or existing geometry rated against changes in process overall duty.

Chemstations™ 11

Heat exchanger heat curve

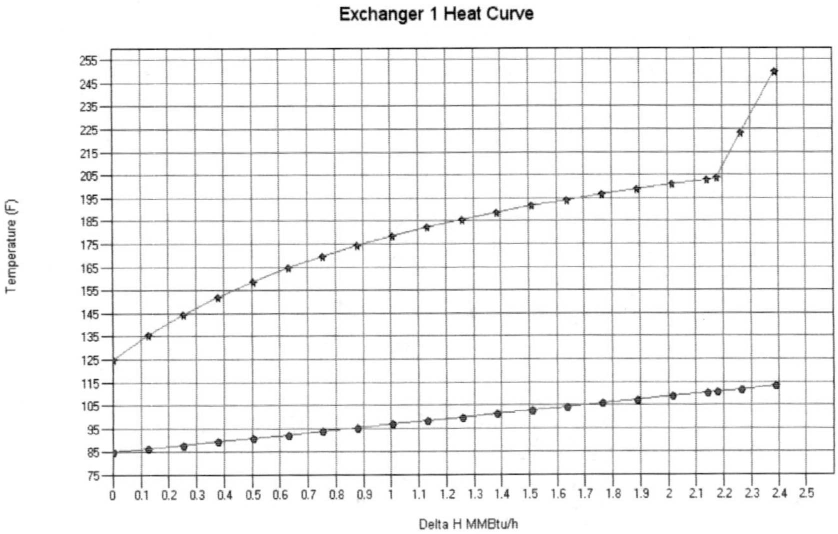

Chemstations™ 12

Heat exchanger process design summary

SUMMARY REPORT

General Data:		Heat Transfer Data:	
Exch Class/Type	R/BEM	Effective Transfer Area	459.63
Shell I.D.	1.27	Area Required	459.62
Shell in Series/Parallel	1/1	COR LMTD	77.06
Number of Tubes	198	U (Calc/Service)	67.55/67.55
Tube Length	12.00	Heat Calc	2.39
Tube O.D./I.D.	0.0625/0.0517	Heat Spec	2.39
		Excess %	0.00
Tube Pattern	TRI60	Foul(S/T)	1.000E-003/1.000E-003
Tube Pitch	0.08	Del P(S/T)	0.50/1.02
Number of Tube Passes	1	SS Film Coeff	118.82
Number of Baffles	5	SS CS Vel	14.61
Baffle Spacing	2.49	TW Resist	0.000202
Baffle Cut, % Diameter	43	TS Film Coeff	304.14
Baffle Type	SSEG	TS Vel	0.91
Baffle space def.	Edge-Edge		

Chemstations™ 13

Heat exchange coefficients as a function of flow rate

- Consider a process where a shell-and-tube heat exchanger has been selected to partially condense a mixture of inert and condensable gases.
- Translate the heat duties and condensation curves from a thermodynamic model into a process design of a shell-and-tube heat exchanger; a heat curve segmenting the entire duty into several segments is generated.
- A preliminary design is established for the design case, and cooling water requirement is calculated.
- Using a heat exchange process design utility, a geometry is selected.

Shell ID Inches	15.24
Tube Count	198
Tube Length Feet	12.0
Tube OD Inches	0.75
Area Sq.Ft.	459.6

How do we test the design for turndown problems?

Chemstations™ 14

Heat exchange coefficients as a function of flow rate

Let the heat exchanger be defined by process design geometry only.

Ask the simulation to calculate the utility rate needed to get the process stream cooled.

The simulator will calculate the overall U, and utility rates and outlet conditions.

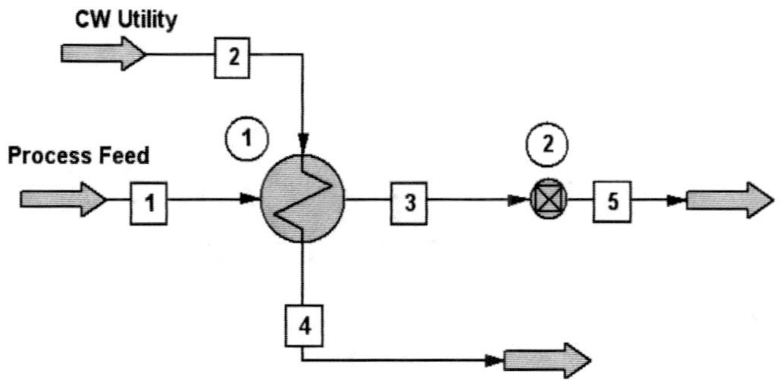

Chemstations™ 15

Summary heat exchanger coefficients

- Red numbers are constrained values; black numbers are calculations.
- Notice that overall U increases with increasing flow as film coefficients are calculated to increase.

Process Flow lbs/hr	12000	14,000	16000	18000
Process Stream T Out	125.1	125.1	125.0	125.1
Process Stream Out VF	0.410	0.411	0.409	0.412
Process Stream dp psi	0.3589	0.5107	0.6579	0.8506
Utility Flow Lbs. Hr	57823	84289.6	129195	207145
Utility Stream dp psi	0.9657	1.0034	1.1923	1.5323
Utility Stream T Out	120.5	113.4	106.2	99.8
U Btu/hr-ft2-F	60.1	67.5	74.6	81.5
Area/shell ft2	460	460	460	460
Calc Ht Duty MMBtu/h	2.0529	2.3923	2.7328	3.0702
LMTD (End points) F	76.2	78.8	81.2	83.3
Process Stream Pout psia	19.6	19.5	19.3	19.1
Utility Stream Pout psia	49.0	49.0	48.8	48.5

Chemstations™ 16

Transport and equilibrium

- VLE usually calculated based on theoretical trays
- "But you can't buy a theoretical tray!"
- You can buy real trays and packing.
- Correlations between real internal performance and theoretical are sparse and/or proprietary.
- Diffusion rate models exist, but are rarely employed on real projects and thus are unproven.

Chemstations™ 17

Transport and equilibrium

- Consider a real plant problem:
  - Existing 20' packed column used to scrub VOC (toluene) out of water using air.
  - If we assume an HETP of 2', the column ought to represent 10 theoretical stages.
  - Use NRTL for water/toluene/inert gas VLLE.
  - Converged column shows toluene content $< 10^{-4}$ pounds 40% down the column.
  - Data from the field[1] suggests aqueous effluent $> 10^{-4}$ pounds.

[1] Private communication with J. Reagan, Process Engineering Associates, LLC

Chemstations™ 18

Transport and equilibrium

- Mass transfer simulation
  - Take converged VLE model and use sizing tools to specify internals. {250Y structured packing}
  - This loads tower internal geometry and performance data from library.
  - Recalculate based on mass transfer limitations.

Chemstations™ 19

Transport and equilibrium

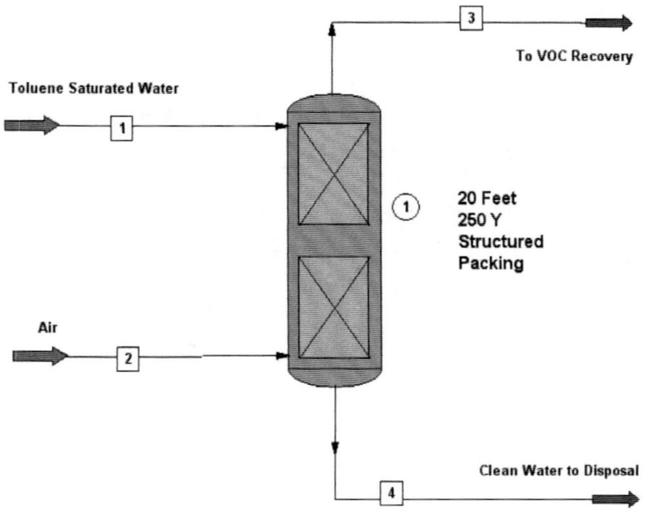

Chemstations™ 20

Transport and equilibrium

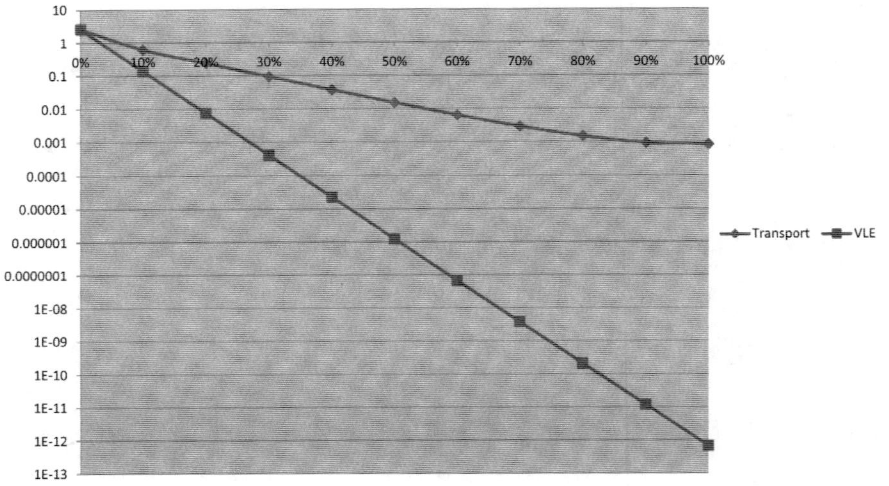

Toluene in Water Phase vs. Column Position

Chemstations™ 21

Transport and equilibrium

- Some observations:
  - Deviations from VLE ideal will be more identifiable when VLE predicts complete removal in one or two stages.
  - Packed columns have a continuous composition gradient, whereas trays can be expected to have discrete contacts.
  - Computers and models have both evolved enough that the testing of transport-based designs against VLE designs should become routine, to see how the models can better reflect reality in the future.

Chemstations™ 22

Conclusions and suggestions

In three examples, we have seen the following:

- Thermo adds insight to a transport problem in a pipe.
- Transport calculations coupled with thermo enable modeling of turndown in a heat exchanger.
- Transport provides an explanation for mismatch between field observations and thermodynamic simulation of a packed column.

1. "If it is truly a mass transfer problem, it should be modeled as one." —J. Reagan.
2. It should never be "one or the other," as insight and simulation output should use every available tool to improve the efficiency and safety of design output.

Chemstations™ 23

A PE Based Industry-Academia Cooperative Chemical Engineering Design Course Heuristics → Excel → Process Modeling

University of Illinois at Chicago, Department of Chemical Engineering
Jeffery P. Perl, PhD, PE, CHMM, Adjunct Professor of Chemical Engineering

Industry Mentors and Student Design II Team Instructors
Shannon Brown, PE and Jerry Palmer, PE – Ambitech Engineers
Bill Keesom and Dennis O'Brien, PE – Jacobs Consultancy
Pat Shannon, PE and Dan Rusinak, PE – Middough Engineers
Adam Kanyuh – UOP
Tom Stephan, PE – Superior Engineering

UIC UNIVERSITY OF ILLINOIS AT CHICAGO

Two-Semester Course Chemical Engineering Design I&II

- Capstone Chemical Engineering Course
- Prepare Students for:
  - Process Design & Engineering Assignments
  - Plant Production & Operations
  - Teaching and R&D

UIC UNIVERSITY OF ILLINOIS AT CHICAGO

Capstone Chemical Engineering Design

Two-Semester Course Chemical Engineering Design I&II

- Fall – 4 semester hours
  - Tools, Tips and Techniques
  - Unit Operations, Simple Examples, ESOH, ECON
  - Quizzes and Exams = 100%
  - Perl + Occasional Guest Lecturer
  - Experienced Based Problems
  - FE and PE Exam Specification as a Guide

UIC UNIVERSITY OF ILLINOIS AT CHICAGO

Capstone Chemical Engineering Design

NCEES Professional Licensure

- NCEES FE/PE ChE Exam Specification
  - Design Specific Elements
  - 19% of the PE Exam
- Practice Based Exam Topics
  - 4 years out of school
- FE taken 7th or 8th semester
  - Design Elements as well

UIC UNIVERSITY OF ILLINOIS AT CHICAGO

A World-Class Education, A World-Class City

FE&PE Exam Spec

Chemical Engineering Design I
University of Illinois - Chicago
ChE 396 Fall 2011
Copyright 2011
Jeffery P. Perl, PhD, PE, CHMM

Aug 23, 2011

Excerpts Pertaining to Chemical Process Engineering Design
National Council of Examiners for Engineers and Surveyors

FROM FUNDAMENTALS ENGINEERING EXAM (FE) (Exam after BS Degree)

Morning ChE FE EXAM
V. Ethics and Business Practices 7%
A. Code of ethics (professional and technical societies)
B. Agreements and contracts
C. Ethical versus legal
D. Professional liability
E. Public protection issues (e.g., licensing boards)

VI. Engineering Economics 8%
A. Discounted cash flow (e.g., equivalence, PW, equivalent annual FW, rate of return)
B. Cost (e.g., incremental, average, sunk, estimating)
C. Analyses (e.g., breakeven, benefit-cost)
D. Uncertainty (e.g., expected value and risk)

VII. Engineering Mechanics (Statics and Dynamics) 10%

VIII. Strength of Materials 7%

IX. Material Properties 7%
A. Properties
1. chemical
2. electrical
3. mechanical
4. physical
B. Corrosion mechanisms and control
C. Materials
1. engineered materials
2. ferrous metals
3. nonferrous metals

AFTERNOON ChE FE EXAM

VIII. Process Design and Economic Optimization 10%
A. Process flow diagrams (PFD)
B. Piping and instrumentation diagrams (P&ID)
C. Scale-up
D. Comparison of economic alternatives (e.g., net present value, discounted cash flow, rate of return)
E. Cost estimation

IX. Computer Usage in Chemical Engineering 5%
A. Numerical methods and concepts (e.g., convergence, tolerance)
B. Spreadsheets for chemical engineering calculations
C. Statistical data analysis

XI. Safety, Health, and Environmental 5%
A. Hazardous properties of materials (e.g., corrosive, flammable, toxic),

UIC UNIVERSITY OF ILLINOIS AT CHICAGO

A World-Class Education, A World-Class City

Fall Semester Highlights

- Prepare for Spring Design Problem
  - Set the Design Basis
  - Draw the Picture
  - Perform a M&E Balance
  - Key Equipment Calculation Procedures
  - Overall Engineering Economics
  - Review ESOH, Sustainability and SAChE
  - Flow Sheeting BFD→PFD→P&ID
  - Conceptual Process Control

UIC UNIVERSITY OF ILLINOIS AT CHICAGO

A World-Class Education, A World-Class City

Fall Semester
Detail

- Design Basis
  - How to set or interpret a problem
- Picture
  - Start with Simple BFD
  - Match to Design Basis
  - Flow Sheeting BFD → PFD → P&ID
- M&E Balance
  - Material balance first
  - Energy balance
  - Coupled M&E Balance for Picture

UIC UNIVERSITY OF ILLINOIS AT CHICAGO

A World-Class Education, A World-Class City

Fall Semester
Detail

- Calculation Procedures
  - Unit Operations and Reactors
    - Heat Exchangers, Pumps, Compressors, Towers, Reactors
- Engineering Economics
  - Time $Value
    - NPV, IRR, Simple Breakeven
- ESOH, Sustainability and SAChE
  - SAChE UIC Graduation Requirement
- Flow Sheeting and Conceptual Process Control
  - Along for the ride

UIC UNIVERSITY OF ILLINOIS AT CHICAGO

A World-Class Education, A World-Class City

Spring Semester
2013 Class Design Problem

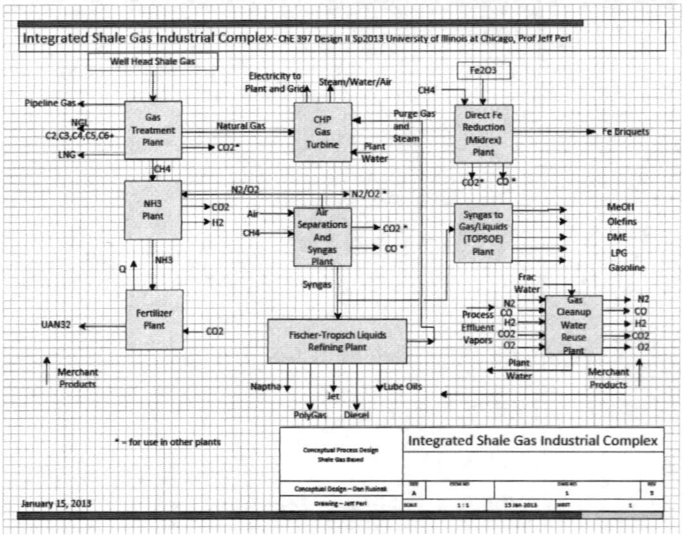

UIC UNIVERSITY OF ILLINOIS AT CHICAGO

Spring Semester
Overview

- Chemical Engineering Design I&II
  - Spring – 3 semester hours
    - Major Industry Design Based Project Exercise
    - Nine industry design experts
    - Nine Student Teams (4 each)
    - Added SAChE Completion for Graduation
    - Students Present in Class, AIChE and UIC/COE
    - Project=75%, Exams=25%
    - FE Exam=Final Exam (80% last year)

UIC UNIVERSITY OF ILLINOIS AT CHICAGO

Capstone Chemical Engineering Design

Spring Semester
Overview

- Excel Timeline/Project Objective Outline
  - Five – 5 hours meetings - Graded
    - 9 Teams talk + Q/A
  - COE Design Expo (Annual for UIC required)
  - AIChE Chicago Poster Competition
- Student Teams – based on first semester

UIC UNIVERSITY OF ILLINOIS AT CHICAGO

A World-Class Education, A World-Class City

Design Calculation Procedures
Sample Heuristic M&E/Design Equations

- Generalized design equations
- Heat Transfer
  - Q=m*Cp*(delta T) →M&E Balance
  - Q=U*A*LMTD →Design sizing
- HP=GPM*(delta P)/(1714*E)
- Applied to Excel spreadsheet
- Excel then used to automate stream
- Aspen to model selected unit

UIC UNIVERSITY OF ILLINOIS AT CHICAGO

Excel MatBal Details

Cells Link Feed with Reactor Rate

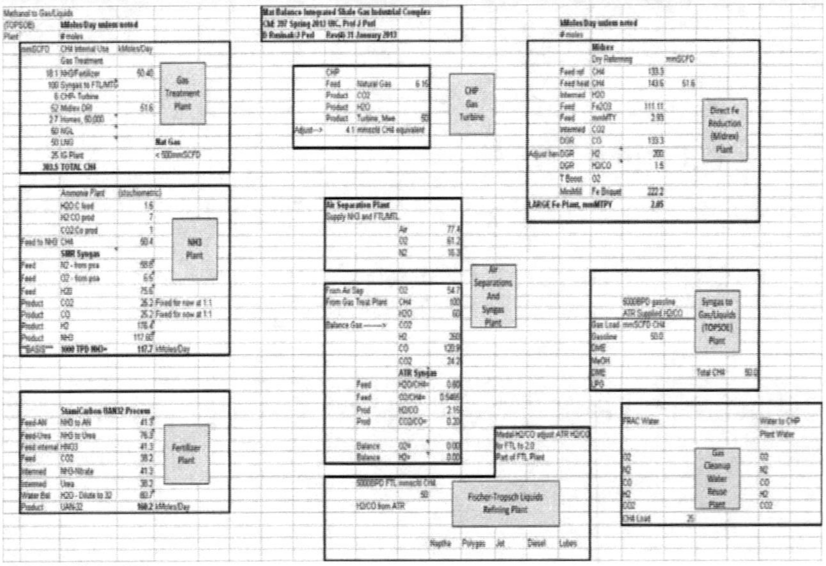

UIC UNIVERSITY OF ILLINOIS AT CHICAGO

Excel MatBal Details

Cells Link Feed with Reactor Rate

Methanol to Gas/Liquids					**Mat Balance Integrated Shale**	
(TOPSOE)		**kMoles/Day unless noted**			**ChE 397 Spring 2013 UIC, Prof**	
Plant		# moles			**D Rusinak/J Perl**	**Rev(3) 11 .**
	mmSCFD	CH4 Intern kMoles/Day				
		Gas Treatment				
	18.1	NH3/Fertili	50.40			CHP
	100	Syngas to FTL/MTG				Feed
	6	CHP- Turbine				Product
	52	Midrex DR	51.6			Product
	2.7	Homes, 50,000				Product
	50	NGL			Adjust--->	4.1
	50	LNG		**Nat Gas**		
	25	IG Plant		< 500mmSCFD		
	303.5	**TOTAL CH4**				
		Ammonia (stochiometric)				
		H2O:C fee	1.5			**Air Separ**
		H2:CO pro	7			Supply NH
		CO2:Co pr	1			
	Feed to NH	CH4	50.4			
		SMR Syngas				
	Feed	N2 - from	58.8			
	Feed	O2 - from	6.5			
	Feed	H20	75.6			From Air S
	Product	CO2	25.2	Fixed for now at 1:1		From Gas

UIC UNIVERSITY OF ILLINOIS AT CHICAGO

Excel MatBal Details

Produced Shale Gas Processing

Block Flow Diagram with Overall Mass Balance

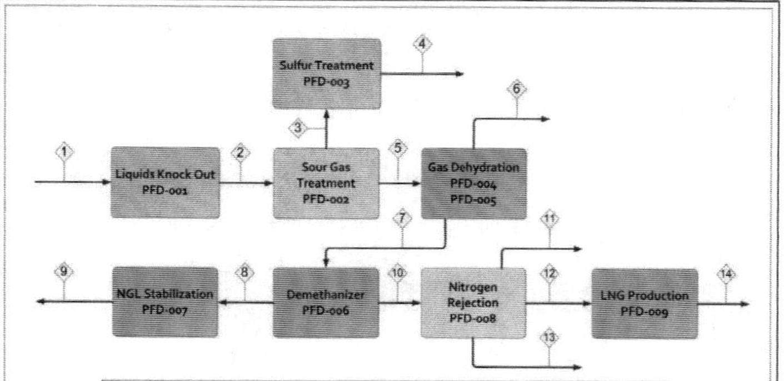

Stream List (LB/HR)

Component	1	2	3	4	5	6	7	8	9	10	11	12	13	14
H2O	184	119	0	0	119	119	0	0	60	0	0	0	0	0
N2	49678	49678	0	0	49678	0	49163	515	63	49163	36402	2076	12631	129
CO2	8539	8533	0	0	8533	8288	245	0	266	237	0	38	196	38
H2S	116	115	115	0	0	0	0	0	1	0	0	0	0	0
H2S (PPM)	100	102	-	0	0	0	0	0	2	0	0	0	0	0
CH4	315293	316271	0	0	315271	0	307791	7544	2911	307750	2990	49586	274076	30684
C2H6	204395	203381	0	0	203381	0	185544	198182	171690	6213	0	1011	5203	1010
C3H8	170370	161376	0	0	161376	0	127537	170370	165341	0	0	0	0	0
C4	75174	57307	0	0	57307	0	33890	75174	74651	0	0	0	0	0
C5	35509	12923	0	0	12923	0	4506	35509	35466	0	0	0	0	0
S (Solid)	0	0	0	85	0	0	0	0	0	0	0	0	0	0
Total	859357	808804	115	0	808567	8407	708875	487283	450450	363362	39392	52712	292109	31861
T (F)	200	78	100	60	78	40	57	0	-21	-82	-280	-188	10	-260
P (PSIA)	815	615	400	-	615	35	600	145	110	488	20	215	120	15

UIC UNIVERSITY OF ILLINOIS AT CHICAGO

Excel MatBal Details

Process Flow Diagram (PFD-001); Liquid Knockout with Initial Control Scheme, Material Balance, and Equipment List

PFD-001 LIQUID KNOCKOUT

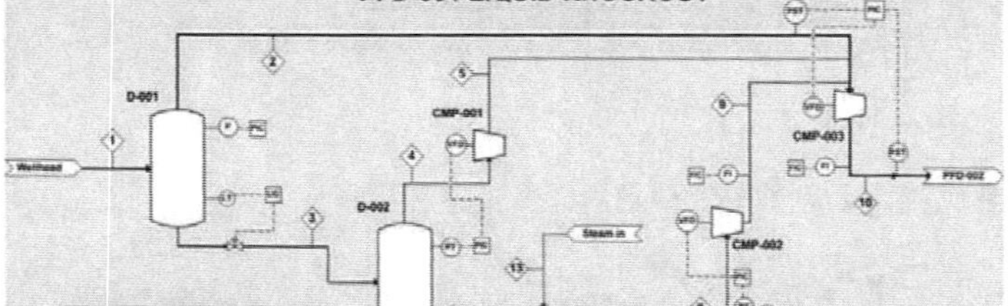

Equipment List

Name	Description
CMP-001	Compressor
CMP-002	Compressor
CMP-003	Compressor
HX-001	Heat Exchanger
D-002	Vapor liquid separator
D-002	Vapor liquid separator
D-003	Vapor liquid separator

Material Balance (LB/HR)

Component	1	2	3	4	5	6	7	8	9	10	11	12	13	14
H2O	184	92	92	8	8	84	84	19	19	119	1	64	0	0
N2	49678	49215	462	432	432	30	30	30	30	49678	0	0	0	0
CO2	8539	8147	392	280	280	112	112	106	106	8533	6	0	0	0
H2S	116.00	104.65	11.35	5.75	5.75	5.60	5.60	4.76	4.76	115.16	0.85	0	0	0
CH4	315293	307977	7316	6152	6152	1164	1164	1141	1141	315271	23	0	0	0
C2	204395	186369	18026	9716	9716	8310	8310	7296	7296	203381	1014	0	0	0
C3	170370	134811	35559	10100	10100	25459	25459	16466	16466	161376	8993	0	0	0
C4	75174	44837	30337	3780	3780	26557	26557	8690	8690	57307	17887	0	0	0
C5+	35509	10193	25316	857	857	24459	24459	1874	1874	12923	22585	0	0	0
Steam	0	0	0	0	0	0	0	0	0	0	0	0	1125	1125
TOTAL	859257	741745	117512	31331	31331	86181	86181	35628	35628	808703	50490	64	1125	1125
T	200.0	59.7	59.7	81.4	184.8	81.4	100.0	40.0	288.7	77.6	40.0	40.0	566	130
P	814.7	600.0	600.0	264.7	600.0	264.7	276.0	44.7	600.0	814.7	44.7	44.7	1200	1200
Vapor Fraction	1.00	1.00	0.00	1.00	1.00	0.00	0.05	1.00	1.00	1.00	0.00	0.00	1.00	0.00

UIC UNIVERSITY OF ILLINOIS AT CHICAGO

Design Calculation Procedures
Excel and Process Models
Automating Calculations

- Aspen, Chemcad, Hysis, etc
  - Estimate Transport Properties
    - Simple energy balances
    - Simple pipe flow
  - Estimate Physical Interaction Parameters
    - Distillation
    - Liquid extraction
    - Scrubbing
    - Compressibility
  - Real-time simulation, operation and control (O'Brien)

Excel and Process Models
Teaching Pitfalls

- Thermodynamic Packages
  - Singularities
  - Incorrect or Missing Package
- Overall Process Model
  - Convergence
    - Multiple steady-states
    - Testing
  - Incomplete Unit Ops or Reactor Model
  - Heat Exchanger cross over (pinch)

Excel and Process Models
Economics Details - General

- Economics – Simple, Heuristics
  - Time value of money
  - Breakeven
  - Net Present Value (NPV)
  - Internal Rate of Return (IRR)
  - Classic Peters & Timmerhaus
  - Classic Grant & Ireson

UIC UNIVERSITY OF ILLINOIS AT CHICAGO

Excel and Process Models
Economics Details

- Economics – Aspen
  - Aspen ICARUS
    - Integrated Process Economics (IPE)
    - Fabricates Equipment
    - Can add total cost evaluation by manual input

- Economics – Excel
  - Functions built in
  - Standard Spreadsheet application

UIC UNIVERSITY OF ILLINOIS AT CHICAGO

Texts

UIC UNIVERSITY OF ILLINOIS AT CHICAGO

Spring 2010 Class Project

UIC UNIVERSITY OF ILLINOIS AT CHICAGO

A World-Class Education, A World-Class City

Punch Line

- UIC Senior Design Course
  - Design, Plant Operations and Academia
- Design Tools
  - Heuristics → Excel → Process Models
- Know the answer before you advance too far
  - Safe, Sustainable Profitable Designs Taught Here
- QUESTIONS?

UIC UNIVERSITY OF ILLINOIS AT CHICAGO

A World-Class Education, A World-Class City

Punch Line

- ABET <u>Accreditation</u>
  - Outcomes oriented
  - Create your program – What constitutes a ChE BS
  - Do it and measure it
- NCEES <u>Examination</u>
  - Developed by team of practitioners and academicians
  - AIChE 25-30 volunteers create and update exam
  - Periodic review/update – continuous improvement
- Balanced Approach – End User Participation

UIC UNIVERSITY OF ILLINOIS AT CHICAGO

A World-Class Education, A World-Class City

Process Development Aspects

Rajat Basu and Jing Guo
Honeywell International
Performance Materials and Technologies

Eddy Yap
Honeywell International
Automation and Control Solutions

Honeywell

Outline

Honeywell

- Project 1 –
  - Steady State Process model of a chemical plant
    - This part will show how a large scale chemical plant in operation for many years can get benefit by having process model built
- Project 2 –
  - Dynamic model of a plant and the development of operator training simulator
    - This part demonstrates how to take the steady state model to a dynamic model and develop operator training simulator to help in better operations

2

Lifecycle for Chemical Process Modeling

Honeywell

Operations Simulation

- Off-Line Training (Simulator)
- Operator Certification
- Error Trapping-Locate & Diagnose
- Upset/Start-Up/Shutdown Simulator

Helps Mfg Eng'rs

Dynamic Simulation

- Optimize Transitions
- Reduce Waste
- Test New Control Systems
- Improve Quality (Reduce Variability)
- Improve Loop Tuning
- Reduce Recovery Time
- Multivariate Control
- Inferential Sensors
- Product by Process

Helps Process Eng'rs

Steady State Simulation

- Design new plants/unit operations
- Reduce Operating Costs
- Increase Capacity
- Reduce Cycle Times
- Define Conditions for New Products
- "What-If Scenarios"
- Improve Quality
- Train Optimization
- Reduce Extractables
- Capture Expert knowledge

Increasing Knowledge and Reach

Impact Depends on Knowledge and Reach

3

Honeywell

Project 1

STEADY STATE PROCESS MODEL OF A CHEMICAL PLANT

4

Objectives of steady state model

Honeywell

- Optimize operating conditions of a large chemical plant across all process areas by leveraging process models
- Support operation team via modeling designed flow schemes
- Global know-how transfer via reliable model package

5

Caprolactam (2-oxohexamethylenimine)

Honeywell

- Structure

- Properties
  - Chemical Formula – C5H11NO
  - Molecular Weight – 113.16
  - Boiling point – 269 °C
  - Melting Point – 69 °C
  - Solubility in water - Soluble

6

Caprolactam Purification Program

Honeywell

- Caprolactam purification:
  - Filtration
  - Re-crystallization
  - Flash distillation
- Major components:
  - Caprolactam
  - Water
- Thermodynamic model, VLE, SLE were developed for caprolactam-water using literature and Honeywell internal data
- For selected impurities: thermo data and binary interactions are predicted

Model Contents:
- *115 Devices*
- *280 Streams*
- *15 Compounds*

7

Plant data vs Model results

Honeywell

Plant data		% relative difference		
		<5%	5-10%	>10%
Mass Flow data #	199	190	3	6
Solid Fraction Data #	17	14	0	3
Temp data #	199	187	5	7
Water content data #	48	41	6	1
Total data comparison #	**463**	**432**	**14**	**17**

Simulation results agree reasonably well with plant data

8

Process Modeling Achievements

Honeywell

- **Phase 1**
  - Thermodynamic models for VLE, SLE model for water caprolactam were developed and validated with literature data
  - Crystallizer *train* baselines were developed and updated and combined with water wash system to have a unified model
- **Phase 2**
  - Crystallizer model validated to show that in steady state condition they reproduce crystal content quite well
  - Individual crystallizer trains have been analyzed, and completed with good agreement
  - Caprolactam entrainment in water vapor incorporated to better match with plant data
  - Significant improvement done in filter model exhibiting how liquid is separated from the crystals

9

Honeywell

Project 2

DEVELOPMENT OF DYNAMIC MODEL AND OPERATOR TRAINING SIMULATOR

10

Model Development

Honeywell

- Thermodynamic Models –
  - Thermophysical properties were measured and correlated and are used to develop plant models right from the design phase
  - Both equation of state and activity coefficient models were developed to be used in various parts of the plant
- Steady State Models
  - Original plant models were developed in ASPEN® and later incorporated into UniSim® process simulator
  - Models were validated on-line with plant data from UNIFORMANCE® data historian
- Dynamic Models
  - Steady state models were converted to dynamic models in UniSim
  - Dynamic models validated with known plant disturbances

11

Development of operator training simulator

Honeywell

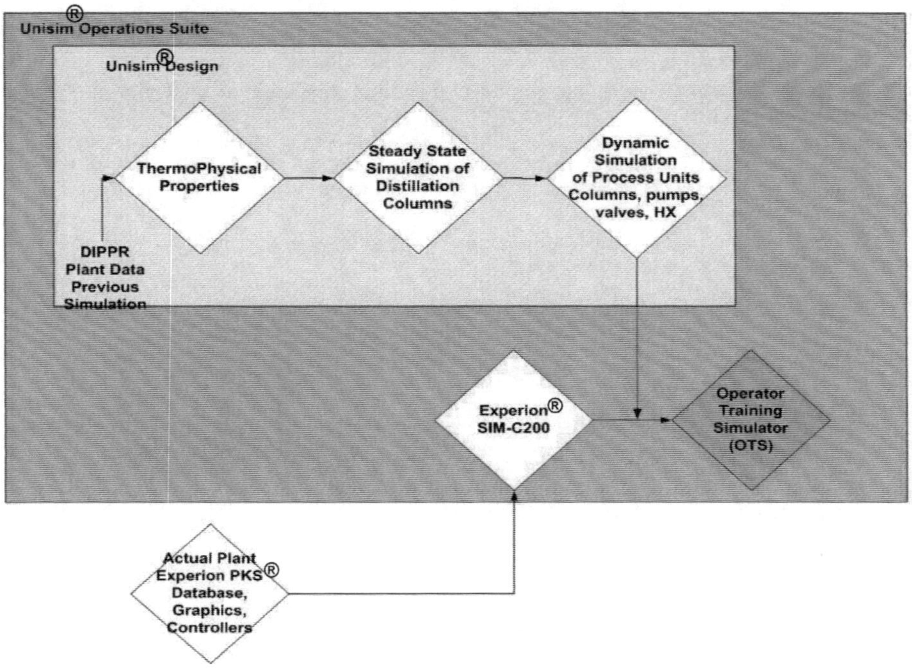

12

UniSim®-Experion® PKS C200 Architecture

Honeywell

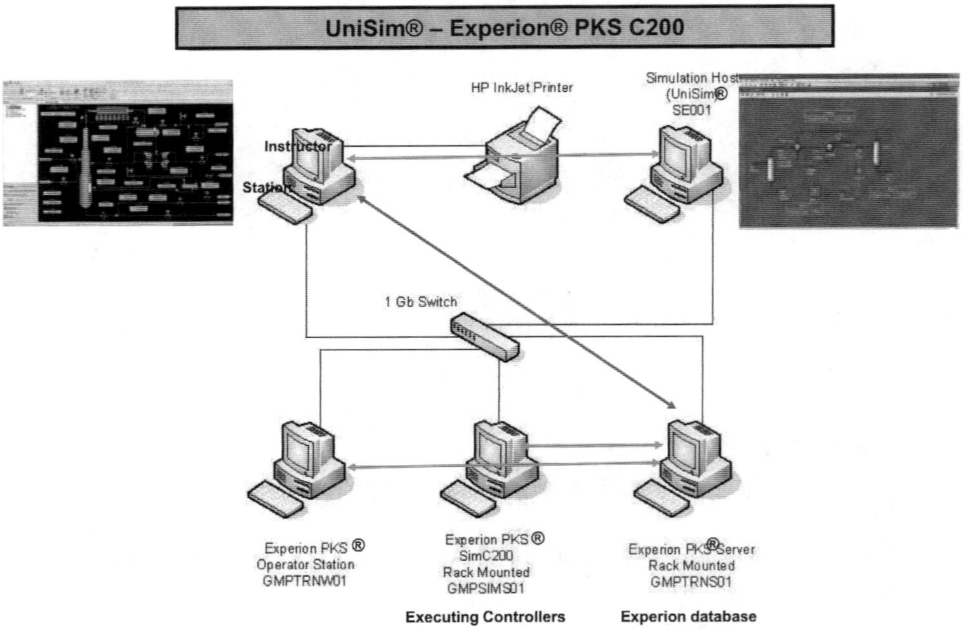

13

Maintenance of OTS

Honeywell

- **Advantages of Virtualization[1]**
  - ➢ **Lower Cost – Multiple Virtual Machines (VM) can run on a single server**
  - ➢ **Greater flexibility – Users can add, remove or change computers and network components at will.**
  - ➢ **Better manageability – Snapshots provide a mechanism to roll back and forth between various backups. This substantially lowers the risk of upgrades and other changes.**
  - ➢ **Easier migration – Data centers can migrate to new hardware by simply copying VMs to a new server.**
  - ➢ **Easier remote support – Easier to duplicate identical system at remote location or easier remote access to the system.**

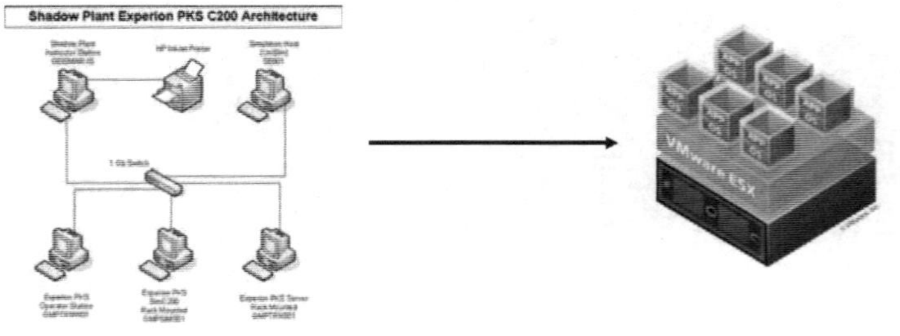

1. Virtualization of Operator Training & Production Systems at Honeywell Specialty Materials by John McIlwain

14

Benefits of OTS

Honeywell

- **Effectiveness and Benefits of OTS has been well documented**[2]
  - **Operator Training System (OTS) applied to Fluorocarbon Plant**
    - First Principles Models built in UniSim® coupled with Experion® System perfectly mimics actual control environment
    - Built-in and custom scenarios train operators for Start-Up, Shut-Down plus abnormal situations
  - **Results:**
    - Reduced time to qualify an operator from >18 months to six months.
    - Reduced start-up time by 20 hours on average.
    - Used OTS in virtual mode to commission new EPKS versions

OTS simulates virtual plants and helps operators

2. The Role of Simulator Technology In Operator Training Programs, Jose Alamo, Nov. 2010 Control Engineering

15

How a Training Simulator Helps Operations

Honeywell

- **Highly Effective Training:**
  - Looks & behaves just like the real plant (Like a Flight Simulator !)
  - No risk to failure
  - Endlessly repeatable
  - Visual: compare performance with best in class.
  - Motivates: fun to do
  - Builds confidence
- **Very Specific & Practical Training:**
  - Address start-up, shut-down, and recurring problems
  - Demonstrate impact of instrument & equipment failures
- **Develop better procedures:**
  - Test improvements before they go on-line
  - Check impact of anticipated problems: power or service failures, etc.
- **Identify and Solve Process Control Issues:**
  - Makes control issues obvious
  - Allows testing of new strategies off-line

16

Operator Training System in Use

Honeywell

Increasing Operator Confidence in Procedures & Best Practices

17

Projected Financial Benefits

Honeywell

- **Estimated Benefits:**
  - Start-Up Speed: Gain 1 day production each product switch; Develop better procedures off-line. Std Work → Improve Std Work
  - Disturbance response improvement - Gain 7-10 hours production each major disturbance
  - Significant $$ benefit payback in about 1-2 years
  - This is in addition to the benefits obtained from steady-state and dynamic simulations used for various plant operations improvement

18

Summary

Honeywell

- This presentation highlights the development, validation of plant models from steady state to dynamic and then to operator training system
- Steady state model of a plant in operation for many years helped optimize productivity and operations
- The advantage of steady state and dynamic in one simulation platform (UniSim®) helped ease the conversion from steady-state to dynamic
- Significant $$ benefit attained by improving operations in the plant

19

Honeywell

Questions

20

Fractionation in Pilot Plants

Dennis O'Brien

Experience

- Pilot Plants in Riverside Development facility
  - 1972
  - Hydrocracking process
  - Fluid Cat Cracking
  - Paraxylene separation
- Heat and material balances for small scale plants in the same facility

Later Professional Involvement

- Fractionation Analysis for many processes
  - Extractive distillation systems
    - Butadiene
    - Aromatics
  - Methanol and Ethanol production

Why Fractionate?

- Need for a recycle stream
    - Fluid Cat Cracking
    - Hydrocracking
- Need to reuse an expensive compound
  - Desorbent systems like:
    - Paraxylene production
    - Normal Paraffin production

Why Fractionate?

- Liquid-Liquid Systems
  - Butadiene extraction
    - NMP
    - Furfural
    - Dimethylformamide
  - Solvents
    - Expensive
    - Dangerous
    - Environmentally harmful

Alternative to fractionation

- Advanced analytic techniques
- E.g.
  - Old- ASTM-D86 Engler
  - New ASTM-D2887 synthetic distillation
    - Number of cut points is greater
    - Detailed analysis by compound is available

	June
% Off	Medium
0	464.3
5	584.8
10	617.9
15	642.1
20	665.1
25	686.3
30	704.4
35	719.9
40	733.2
45	744.8
50	755.4
55	765.4
60	775.2
65	784.8

			June
Component	Weight	Volume	Mole
Methane	<0.01	<0.01	<0.01
Ethene	<0.01	<0.01	<0.01
Ethane	<0.01	<0.01	<0.01
Propane	0.56	0.79	1.21
Isobutane	1.56	1.98	2.54
Isobutene	<0.01	<0.01	<0.01
1-Butene	<0.01	<0.01	<0.01
1,3-Butadiene	<0.01	<0.01	<0.01
n-Butane	6.08	7.42	9.91
trans-2-Butene	<0.01	<0.01	<0.01
2,2-Dimethylpropane	0.01	0.01	0.01
cis-2-Butene	<0.01	<0.01	<0.01
1,2-Butadiene	<0.01	<0.01	<0.01
3-Methyl-1-butene	<0.01	<0.01	<0.01
Isopentane	3.6	4.1	4.72

This goes on for 440 compounds

Heat Sensitive Compounds

- Wiped film distillation stages
  - Available from several vendors.
- Short path distillation
  - Sigma-Aldrich
  - Incon process systems
  - And others

Molecular Distillation

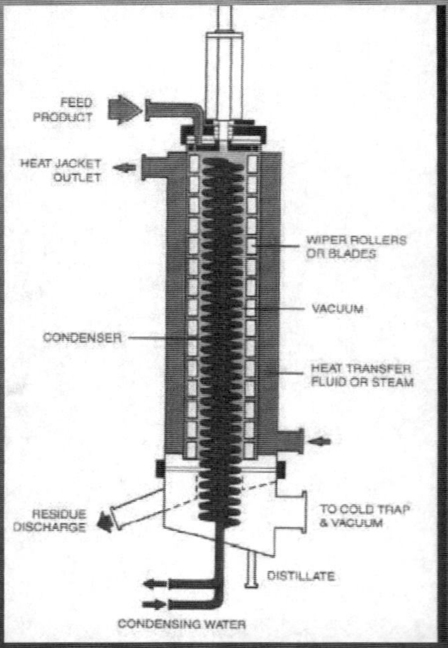

References

- Brazilian Journal of Chemical Engineering : Hetp evaluation of structured packing distillation column Orlando, A.E.Jr.; Medina, L.C.; Mendes, M.F.; Nicolaiewsky, E.M.A.
- Handbook of laboratory distillation : with an introduction into the pilot plant distillation Erich Krell

LIST
DRY PROCESSING - INTELLIGENT PROCESSING

2013 AIChE Process Development Symposium

Heat Transfer in Multiphase Processes using Kneader Reactors

Boyd T. Safrit, PhD, PE

LIST
LEADER IN HIGH VISCOSITY PROCESSING TECHNOLOGY

June 12, 2013

Introduction

- Intro to kneader reactor technology
- Modes of heat/energy transfer
- Rheological Phases
  - Liquid
  - Free flowing solids
  - Pasty
- Integrated modes of heat/energy transfer
  - Liquid → pasty → solids
  - Main evaporation
  - Bulk polymerization with evaporative cooling

June 12, 2013 p. 2

Kneader Technology

- Excellent mixing and kneading
- Large heat exchange area
- Effective self-cleaning
- Efficient renewal of phase boundary layers
- Large working volumes
- High torque applications

June 12, 2013 - p. 3

Modes of heat/energy transfer

- Contact surface area
  - Temperature
  - Viscosity
  - Density
  - Geometry
  - Shaft speed
  - Surface area/volume ↓ as volume ↑
- Mechanical energy input
  - Viscosity
  - Shaft speed
  - Shaft design
  - Scales with volume

June 12, 2013 p. 4

Rheological Phases

- Liquid
- Free flowing solids
- Pasty

Liquid Phase

- Contact area transfer only
- Heat transfer fluid (HTM)
- h_o - local HTM to wall coefficients
- h_i - local process to wall coefficient
- U - overall coefficient
- Energy transfer in/out
  - $\Delta T = T_{HTM} - T_{process}$

Limiting Factor

$$\frac{1}{U} = \frac{1}{h_o} + \frac{\Delta x}{k_{metal}} + \frac{1}{h_i}$$

Free Flowing Solids Phase

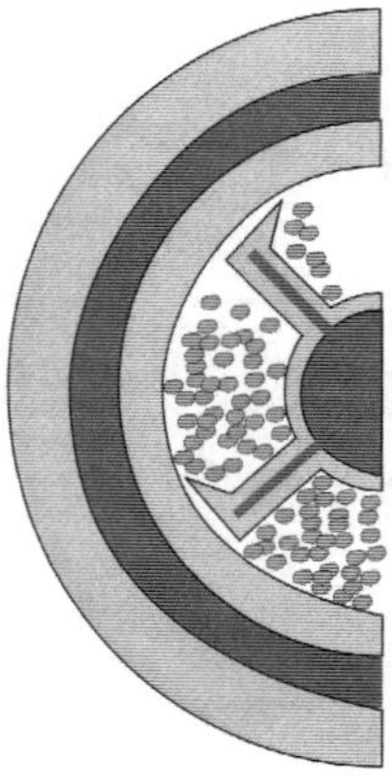

- Contact area transfer only
- Less process contact with walls
- Energy transfer in/out
  - $\Delta T = T_{HTM} - T_{process}$

Limiting Factor

$$\frac{1}{U} = \frac{1}{h_o} + \frac{\Delta x}{k_{metal}} + \frac{1}{h_i}$$

Free Flowing Solids Phase

- Alternate energy transfer model
  - Drying applications

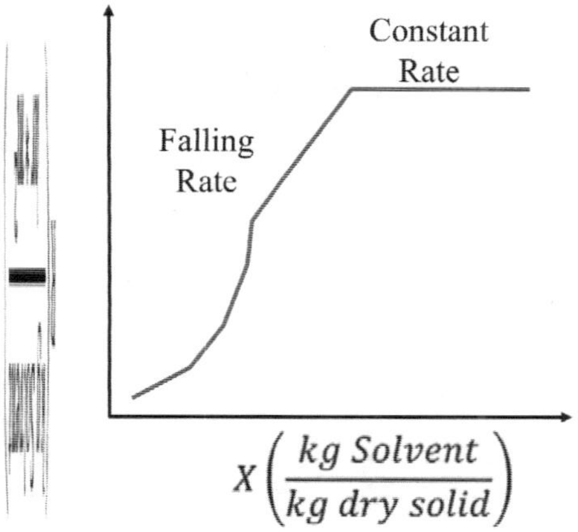

$$R = -\frac{M_{Solid}}{A}\frac{dX}{dt}$$

$$R_C = \frac{U(T - T_W)}{H_{vap}}$$

Geankoplis, 1983

Pasty Phase

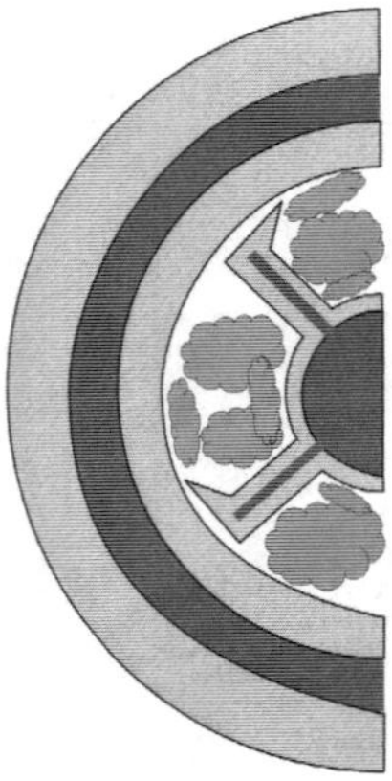

- Contact area transfer in/out
  - $\Delta T = T_{HTM} - T_{process}$
- Mechanical energy transfer in
  - Torque measured by hydraulic pressure
  - Power = Torque * speed
  - Scales with volume

Integrated Transfer Modes

- Liquid → pasty → solids
- Main evaporation
- Bulk polymerization with evaporative cooling

Integrated Transfer Modes

Liquid - Pasty - Solids

- 20-30% solids in organic solvent
- Product > 95% solids

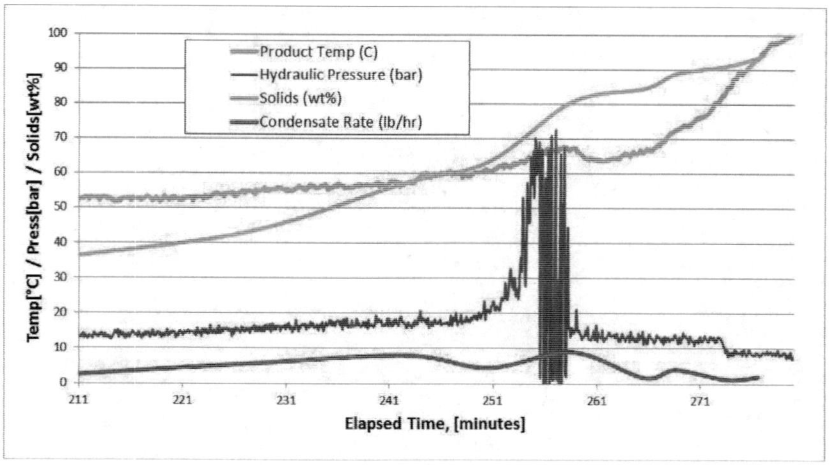

Integrated Transfer Modes

Liquid - Pasty - Solids

- R_c = 13.8 kg solvent/(hr*m^2)
- $R_{falling} = A{\cdot}X+B$

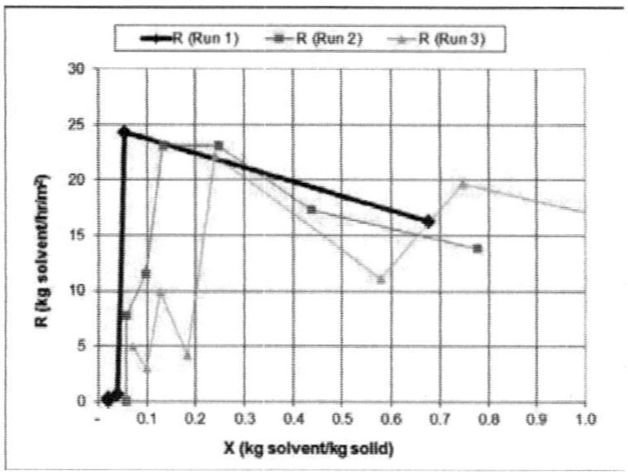

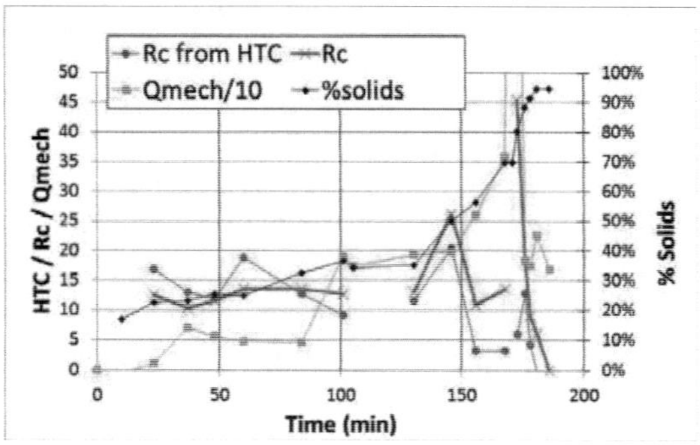

Integrated Transfer Modes
Liquid – Pasty – Solids

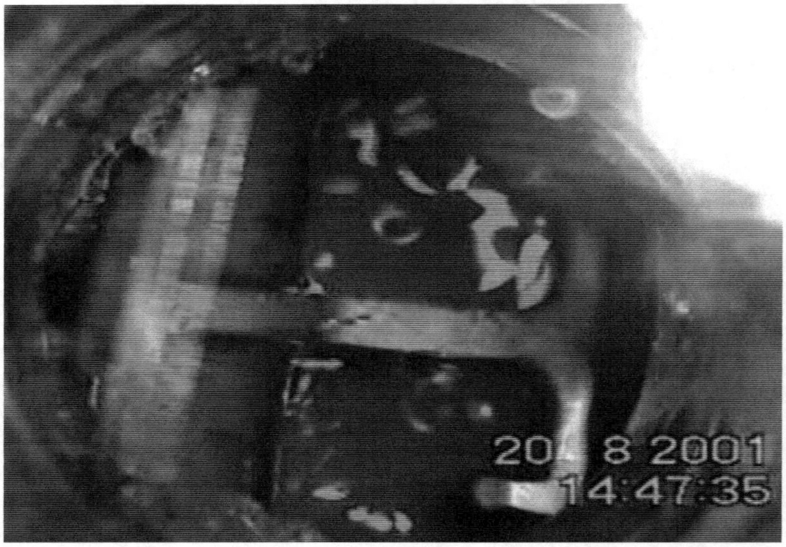

Integrated Transfer Modes
Main Evaporation

- Temperature sensitive elastomer solution
- Feed: 15-20% BR in hexane
- Maximum temperature of 100-130°C
- Discharge target of 2-10% solvent
- High energy duty for solvent evaporation
- Followed by finishing kneader reactor
  - <100 ppm solvent

Main Evaporation

Temperature/Viscosity Profile

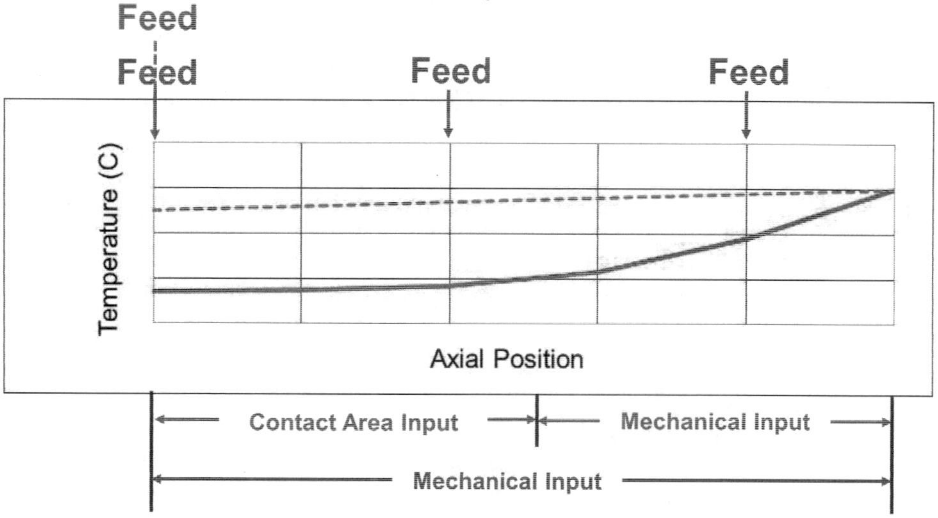

Integrated Transfer Modes

Main Evaporation

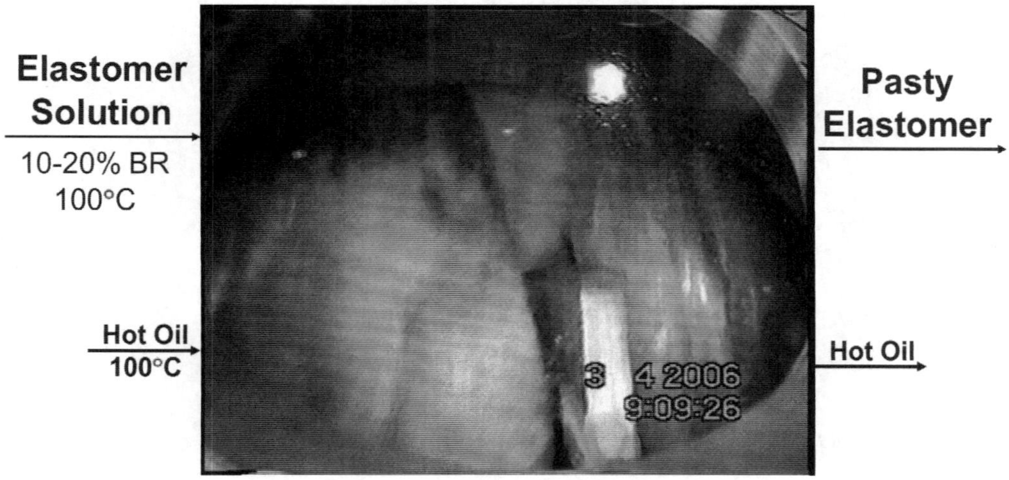

Integrated Transfer Modes

Bulk polymerization with evaporative cooling

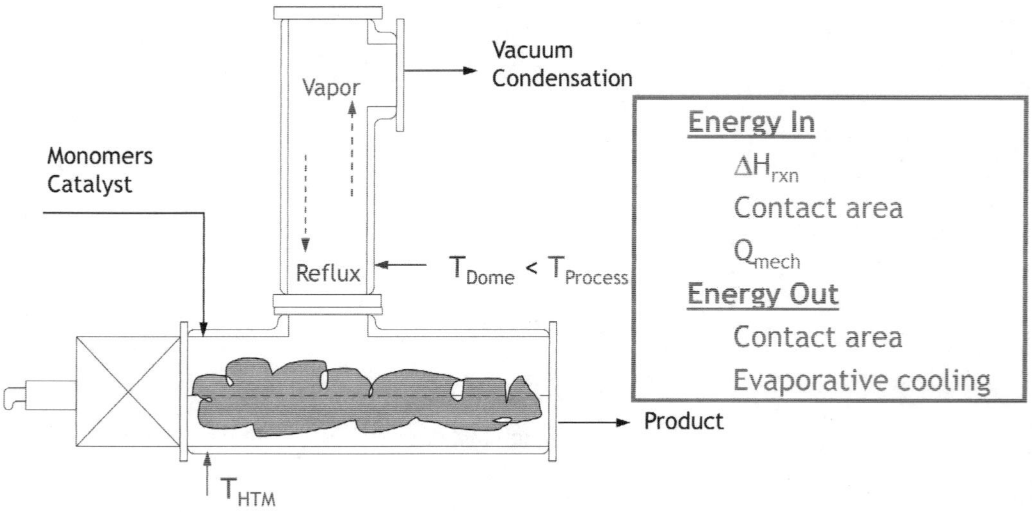

Conclusions

- Kneader reactor technology
  - Flexible
  - High viscosity applications
  - Multiple rheological phases
  - Handles phase transitions
- Heat/energy transfer
  - Challenge to scaleup
  - Multiple modes
  - Integrated modes

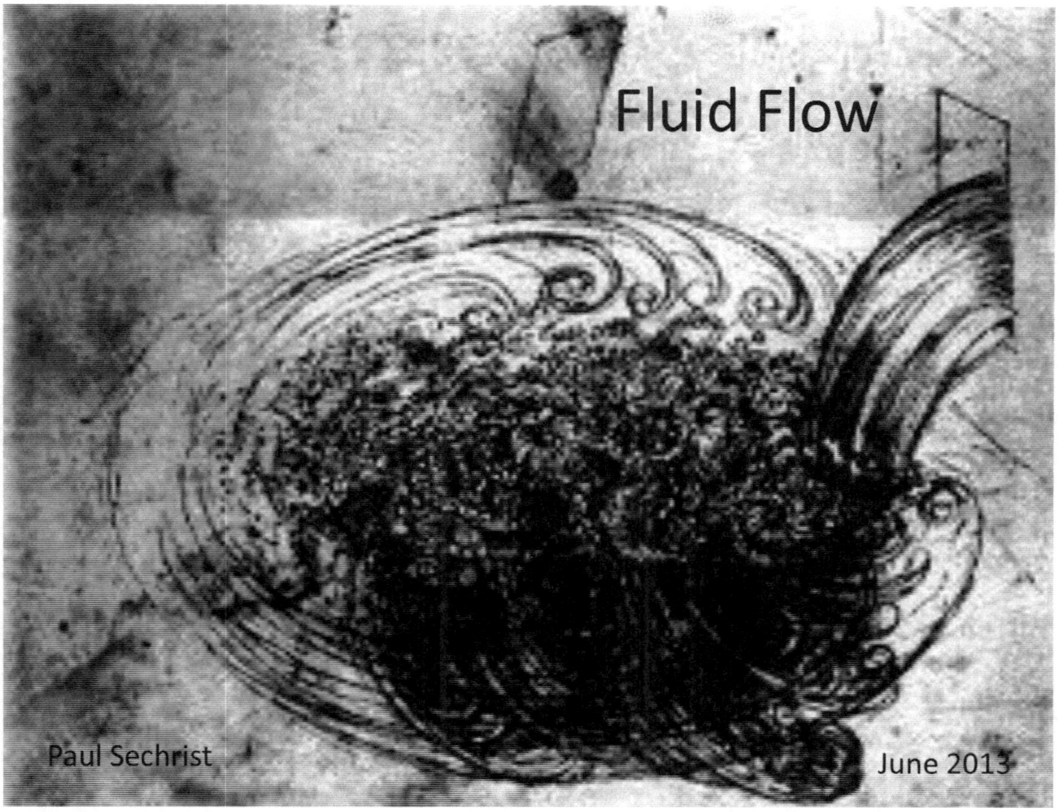

Focus on what's important

- You can't match everything at small scale
- Run simple experiments, then add complexity if needed (recycles, contaminants, etc)
- You can't match reactions with high heat effects
  - Exothermic reactions will be more stable at small scale
  - Endothermic reactions will be faster at small scale

Important Fluid Properties

- Can a second phase form during processing?
- Are there trace components that can react, e.g. during compression?
- Are the bulk properties well-known and available in comm'l simulators
- Is the fluid Newtonian?
- What are the expected ranges of properties?

Dimensionless Numbers

• Reynolds	inertial/viscous
• Froude	inertial/gravity
• Prandtl (Schmidt)	viscous/thermal(conc.)
• Nusselt	inertial/thermal(conc.)
• Weber	inertial/surface tension

Vortex Shedding

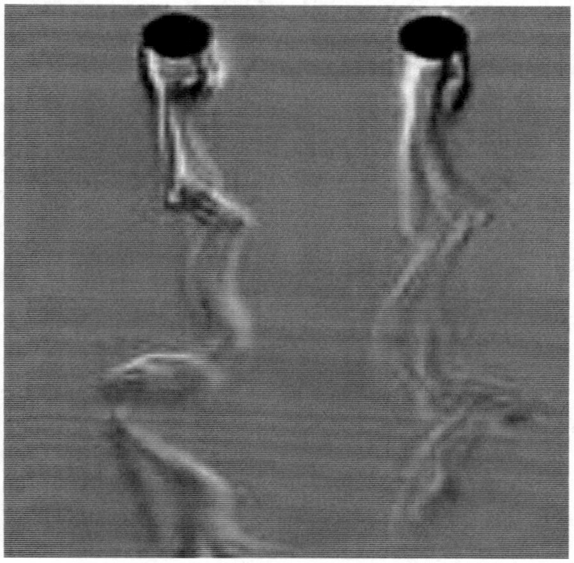

Pressure drop an important tool

Typical particulate bed

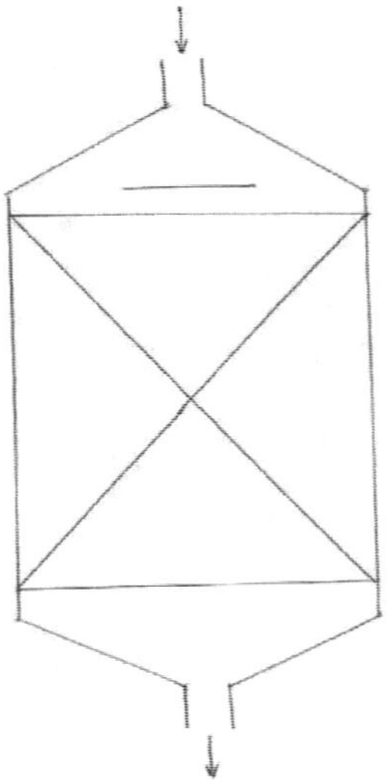

Inlet momentum a problem

- Can disturb bed
- Can cause mal-distribution
- Inlet baffle helps
- Beware of making eddies
- Full width distributor
- Diffuser an option
- Outlet not important if radial dP small

Pressure Drop thru Packed Bed

- Ergun Equation for single phase thru packed bed
- Tosun correction for 2-phase flow thru packed bed (Ind. Eng. Chem. Process Des. Dev., 1984, 23, 35-39)
- Larkin Equation

Hydrodynamic Instabilities

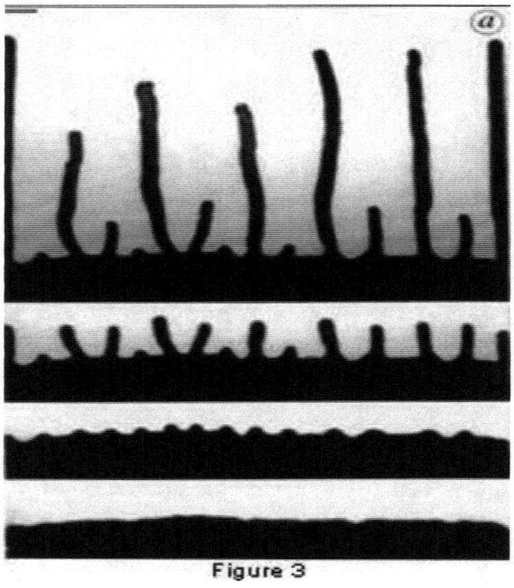
Figure 3

- Can occur when dP low
- Viscosity-driven (Safman-Taylor)
- Density driven (composition or temperature)
- Can reduce by adding dP, or internal dP zones

Have a safe trip home!

Scale-Up Droplet to Production Spray Dryer

Turner, N., GEA Process Engineering Inc.

Originally presented on: 6/12/2013 14:00:00 - 14:30:00

Spray Drying has been the chosen technology for discrete particle production dating back to the early 1900's. Computational Fluid Dynamics (CFD) is one of the developments utilized to better understand the behaviors of various products processed in spray dryers. With the extensive research using CFD there was a gap regarding the effects of heated gas on the atomized droplets in the chamber. Recently this gap was filled by the development of the Drying Kinetics Analyzer (DKA). The DKA opens the doors to visually evaluate the drying process of a single droplet.

Now with the data gathered in DKA analysis can be utilized with CFD modeling in designing new spray drying processes and improve existing installations.

In this presentation we will discuss our ability to examine how an individual particle dries and how varying conditions can impact the final powder morphology and properties. This allows the spray dryer to produce a "custom particle" through adjustment of spray drying parameters determined by DKA. For example, some might want a strong particle not easily damaged during transport; others might require a lightweight powder that dissolves easily. Use of DKA can determine if a given feed will dry into hollow spheres, shriveled particles, "doughnut" shape, etc., and how to design the drying process and equipment to attain a certain type of particle.

Commercial Evaporator Types – An Overview

John F. Micheli, P.E.
Middough, Inc.

Definition

- An evaporator is a separation operation that uses heat transfer to remove or boil off a solvent from a solution.
- The solute has an very low or insignificant vapor pressure compared to the solvent. In most applications, the solvent is water.
- The goal may be to concentrate a solution or to condense and collect the vapor.
- Northeast Illinois is the evaporator & crystallizer capital of North America.
  - Dedert Corporation – Homewood, IL
  - Swenson Technology – Monee, IL
  - HPD/Veolia Water Solutions & Technologies – Plainfield, IL
  - GEA Process Engineering (crystallization) – Hinsdale, IL
  - InCon Processing/Separation Technologies – Batavia, IL
- My personal pilot plant nightmare at the right that started me off.

(glass rising-film evaporator in the Chemical Engineering Unit Operations lab at the Illinois Institute of Technology)

Basic Theory of Operation

- An evaporator has two basic components:
  - A heat exchanger to add heat or remove heat from the system
  - An expansion chamber for allowing the vapor to flash and separate from the heated concentrated liquid or liquor.
- Heat added for evaporation is defined by $Q = U*A*\Delta T$
  - Where Q = heat transfer rate (BTU/hr)
  - A = heat transfer area (ft^2)
  - ΔT = temperature difference (^0F)
  - U = overall heat transfer coefficient - HTC (BTU/(hr* ft^{2*} ^0F))
- HTCs for 2-phase flow in falling-film & rising-film evaporators are determined by testing or operating experience.
- Heat transfer is established by a temperature frame accounting for the boiling point elevation of a concentrated solution or liquor.

Considerations for Evaporator Design

- The operating parameters need to be specified:
  - Capacity
  - Fluid properties (see next slide)
  - Available utilities
  - Control, automation
  - Feed & product concentrations
  - Temperatures
  - Annual operating hours
- The design needs to be economical & account for the following:
  - Optimum heat transfer conditions using the least amount of heat transfer surface
  - Efficient separation of the vapor from the liquor
  - Achieving the product requirements
  - Using the least amount of energy for the design
  - Minimizing fouling of the process & heat transfer surfaces
  - Site conditions - Available space, volume, & headroom for the installation
  - Material of construction (evaporator vendors require the user select the MOC)
  - Robust & safe design for the operating pressures & temperatures (usually ASME Section VIII, Division 1)
  - Proper control anticipating startup, normal, & upset operating conditions to "live" in the overall process

Considerations for Evaporator Design

- Testing or operating experience may be necessary to determine the fluid properties and any potential problems.

- **Solids Composition** - What portion of the solids are dissolved or suspended.
- **% Solids vs Boiling Point Elevation** - The solution boiling point will increase as the solution concentration increases. The boiling point elevation is more pronounced for salt solutions than organic-based solutions.
- **% Solids vs Density** - The density will increase as the solution concentration increases.
- **% Solids vs Solubility** - Solubility usually increases as the solution temperature increases. Solutions with inverse solubility will come out of solution at higher temperatures (fouling).
- **% Solids, Temperature vs Viscosity** - The solution viscosity will usually decrease as the solution temperature increases and increase as the solution concentration increases.
- **% Solids vs Heat of Concentration/Dilution** – Additional heat required for strongly ionic solutions as the solution concentration increases e.g. sulfuric acid, caustic.
- **pH & Corrosivity** – The materials of construction selected need to be compatible for the liquor concentrations & temperatures.
- **Other solution properties**
  - Tendency to foam
  - Tendency to foul
  - Temperature sensitivity
  - Suspended solids properties e.g. particle size, shape, abrasiveness

Basic Components

- A heat exchanger for adding heat to the process. These are usually called heating elements, steam chests, or calandrias.
- An expansion chamber or vapor separator for allowing vapor to boil or flash from the surface & separate the vapor from the liquid.
- The flashing liquid will generate droplets of various sizes. Small droplets may be carried with the vapor which depends on the size & weight of the droplets, vapor velocity & density, and surface tension.
- The vapor separator is designed for a sufficiently low vapor velocity to allow large droplets from the flashing liquor to settle by gravity. Smaller droplets are captured by a mechanical entrainment separator to capture or coalesce most of the droplets for recovery e.g. de-misting wire mesh, centrifugal vanes, chevron plate separators
- Vapors are reused or condensed.
- Evaporation may occur using sub-atmospheric or positive pressures. The system shown is designed for vacuum evaporation.

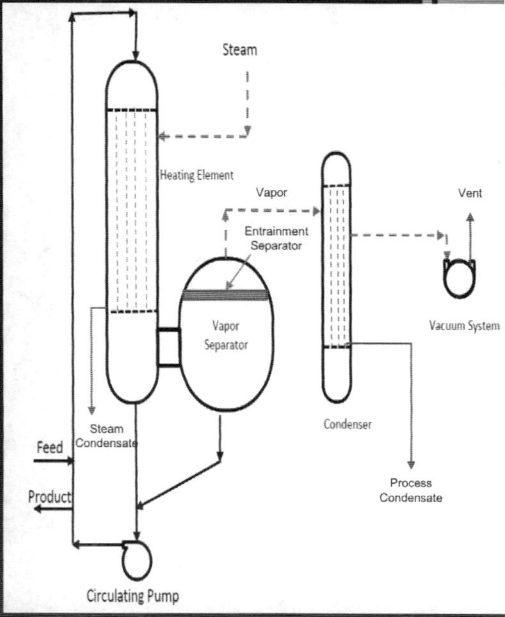

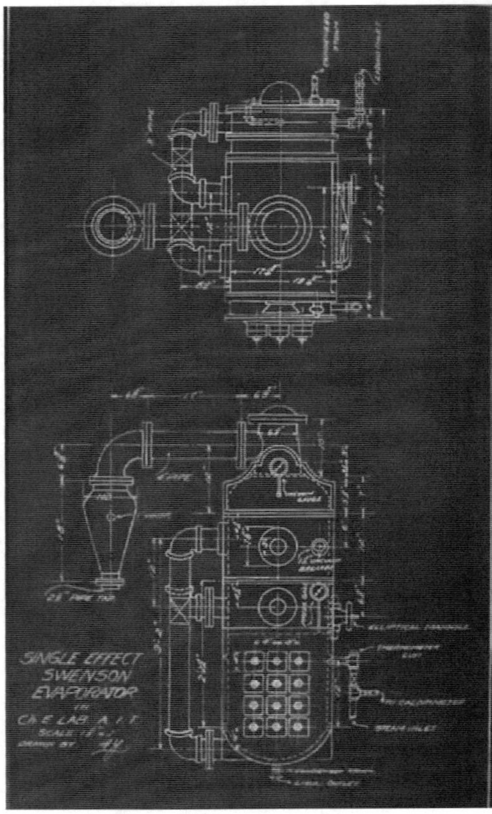

Horizontal Tube Evaporator

- Old design characterized by low headroom
- Steam condenses inside the tubes.
  Liquor contacts the outer tube surfaces by liquid level or by sprays positioned over the tubes
- Horizontal tube units were suitable for low-fouling liquors and where the liquor solids remained in solution
- The design at the left was a laboratory evaporator installed at the Armour Institute of Technology (now Illinois Institute of Technology) and was included in a BSChE thesis by two students in 1915 "Heat Transmission in a Vacuum Evaporator"
- A modern versions of an HTE below is the Aqua Chem vapor compression used for desalinization & water purification.

Short-Tube Vertical Evaporator (Calandria Evaporator)

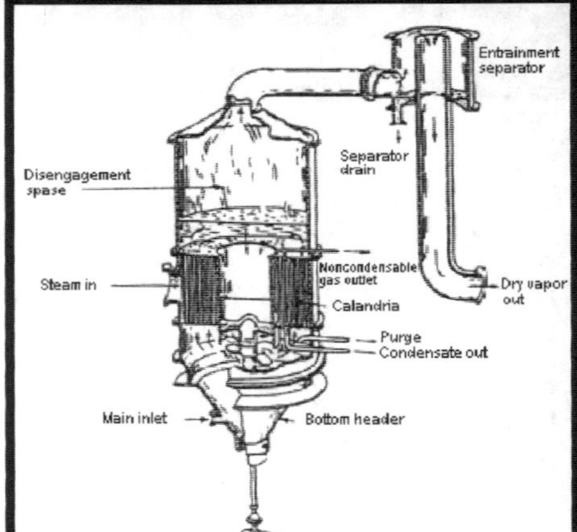

(Drawings from Thermopedia)

- Feed enters at the lower chamber & heats to the boiling point as it rises up the tubes.
  Two-phase flow of concentrated liquor & vapor exits the top of the tubes. The central down-comer allows liquor to recirculate to the bottom of the tubes.
- This is one of the oldest evaporator types .
- **Advantages**
  - Easy to control & operate
  - Insensitive to minor process upsets (thermal inertia)
  - Due the tubesheet design, the body cross section is usually large enough for gravity settling of entrained droplets
  - Good for low head room situations
- **Disadvantages**
  - HTC is dependent on the liquor level.
  - Shorter tubes results in more tubes for the required heat transfer surface and may be expensive to manufacture due to the tubesheet design. Tubesheet may be thick.
  - High residence times
  - If the condensed vapors require low solids content, a demister pad or device may still be required.
- **Typical Process Applications**
  - Sugar syrup concentration, low-level radioactive waste

Rising Film Evaporator - Long-Tube Vertical (RF-LTV)

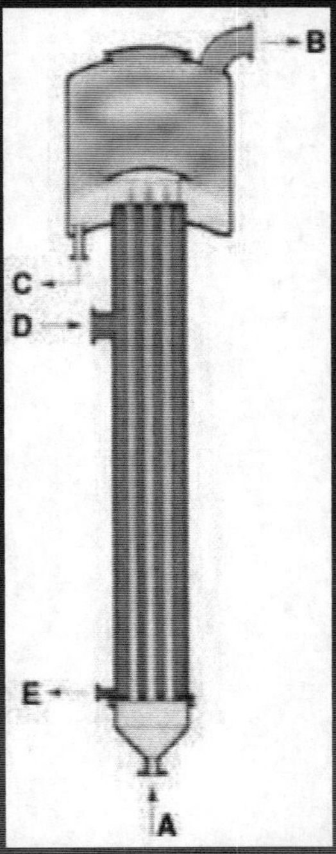

- Feed enters at the lower chamber at A, heats to the boiling point, and then boils as it rises up the tubes forming an annular two-phase flow of concentrated liquor & vapor that exits the top of the tubes. Concentrate exits at C & vapor exits at B.
- If the liquid is cold, it will form a non-boiling zone as it rises where the liquid is heated by conduction.
- When the liquid achieves the boiling point, it percolates up through the tubes. & convective heat transfer develops. Concentrated liquid or liquor moves up by the expanding vapor & forming a thin film in the upper portion of the tubes.
- A jet containing vapor & liquor exits the top of each tube. An impingement plate above the tubes stops the jets and allows the liquor to disengage from the vapor.

  A: Product B: Vapor C: Concentrate

  D: Heating Steam E: Condensate

(Sketch from © GEA Process Engineering Inc." - Evaporator Type)

Rising Film Evaporator - Long-Tube Vertical (RF-LTV)

- **Advantages**
  - Simple construction
  - Much less expensive than comparable short-tube evaporator designs
  - No special feed distribution to the tubes
- **Disadvantages**
  - Gravity will slow the upward movement of the liquor.
  - The ΔT needs to be high enough to create effective 2-phase flow out the tubes.
  - High residence times
  - HTCs have to be determined by experience or testing
- **Typical Process Applications**
  - Paper mill liquors (black, green, white), gelatins, sugar syrups, non-crystallizing inorganic solutions

Falling Film Evaporator (FF-LTV)

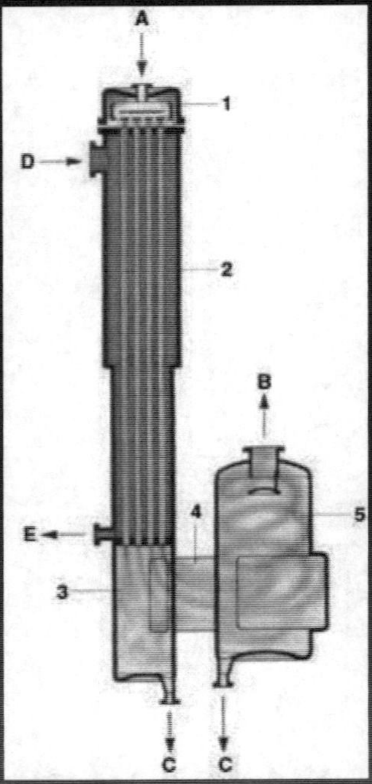

(Sketch from © GEA Process Engineering Inc." - Evaporator Type)

- Feed enters at the upper chamber at A & is distributed to the tube openings. The liquor wets the inner tube surfaces & heats to the boiling point as it drops.
- Two-phase flow of concentrated liquor & vapor exits the bottom of the tubes forming a thin film. Concentrate exits at C & vapor exits at B.
- Co-current vapor & liquor flow design is the most common.

A: Product	1: Head
B: Vapor	2:Heating Element
C: Concentrate	3. Heating Element, lower part
D: Heating Steam	4: Vapor Duct
E: Condensate	5: Vapor Separator

- **Typical Process Applications**
  - Corn syrups, dairy, pharmaceutical, fermentation broths, temperature-sensitive feeds, paper mill liquors, chemicals, mechanical vapor recompression applications

Falling Film Evaporator (FF-LTV)

- **Advantages**
  - Hold up & residence times are low
  - No hydrostatic head to affect the boiling temperature
  - Good for temperature sensitive liquors
  - Lower ΔT needed for heat transfer since gravity helps to pull the liquor down
  - Lower pressure drop on the process side
  - Higher HTCs than rising-film evaporators
  - The vapor body may be side-mounted or bottom-mounted on the heater
- **Disadvantages**
  - More expensive than rising-film evaporators
  - The liquor flow **A** needs to be high enough to wet all of the inner tube surfaces
    - Liquor may need to be recirculated for adequate tube wetting
    - May need a liquor distribution feed system for cold non-flashing feeds
    - Feed pre-heating may be required to flash & distribute the feed
    - Requires pumps to transfer the feeds to the upper chambers
  - HTCs have to be determined by experience or testing
  - Viscosity limited (by evaporator supplier criteria)
  - High headroom requirements

Rising-Film/Falling-Film Evaporator

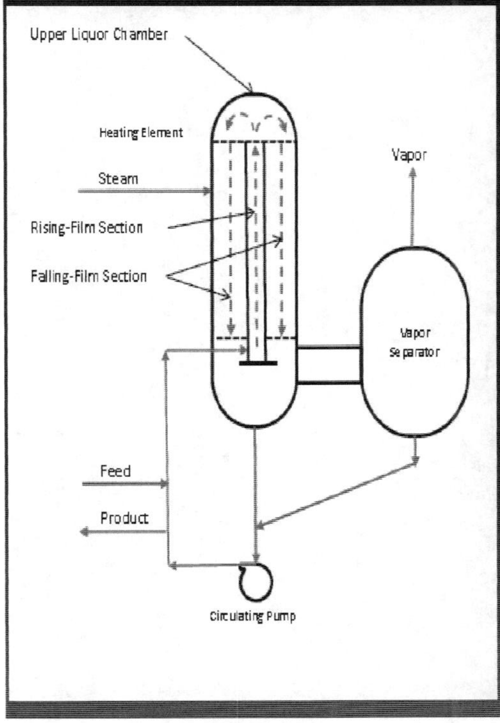

- A hybrid of a rising film & falling film evaporator
- A portion of the tubes at the lower tubesheet are segregated for the rising film section
- The rising film section heats the liquor to the boiling point.
- Liquid & vapor escapes the tubes. The flashing liquid froths or ebulates* over the tube openings to the falling film section.
- **Advantages**
  - The rising film section acts as an internal pre-heater
  - The flashing liquor at the top will help to distribute liquid to all of the open falling film tubes rising film section is symmetrically placed with respect to the falling film tubes.
  - The need for mechanical liquor distribution to the top tube openings is eliminated or minimized.
  - External piping is minimized & more accessible compared to feeding at the top.

* Credit to the late Bill Dedert for this description

Forced Circulation Evaporator (FC)

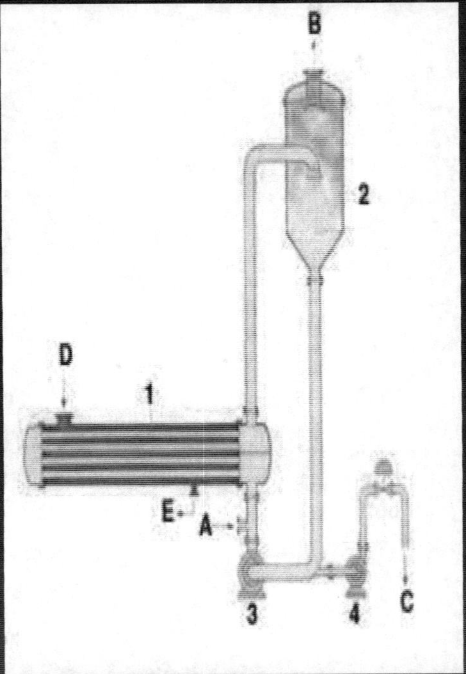

(Sketch from © GEA Process Engineering Inc.® - Evaporator Type)

- Feed enters at the circulating piping at A and enters the heater. A large quantity of fluid circulates through the heater to provide a constant tube velocity. Boiling can be suppressed until it enters the separator & becomes two-phase flow.

A: Feed	1) Heat Exchanger
B: Vapor	2) Flash Vessel (Separator)
C: Concentrate	3) Circulation Pump
D: Heating System	4) Concentrate Pump
E: Condensate	

- The temperature rise through the tubes is controlled (usually 2-10°F depending on the application). The boiling may be suppressed in the heater by hydrostatic head above the liquor outlet or by a restriction in the line before the separator
- The sketch shows the liquor flashing into the separator. Other designs may have the heated recirculated liquor entering below the liquor level to release the flash vapors across the entire separator cross section. A tangential submerged inlet would create a swirling liquid for thermal & physical mixing.

Forced Circulation Evaporator (FC)

- Advantages:
  - Good for scaling or fouling liquors
  - Better for high solids or viscous liquors than LTV FF or RF
  - Good for high suspended solids or crystallizing service
  - Can predict HTCs if the fluid properties are known
  - The heating element may be a horizontal multi-pass as in the sketch or a vertical single or multi-pass (see MVR sketch)
- Disadvantages:
  - Expensive
  - Horsepower intensive due to circulating pump
  - Long residence time
  - Prone to foaming with certain liquors
  - High headroom & space requirements
  - Flashing inlets prone to solids buildup & high entrainment
  - Submerged inlets require adequate level control to avoid splashing from vapor release
- Typical Process Applications
  - Crystallizing or salting services, high-solids liquors (limited by end viscosity), finishing step after a multiple-effect train

Plate Evaporator

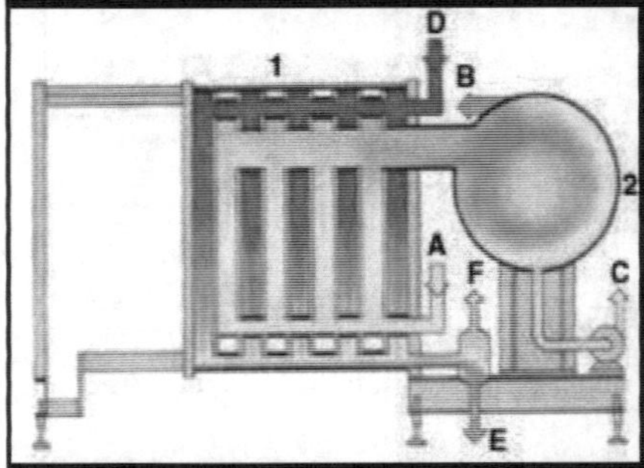

(Sketch from © GEA Process Engineering Inc." - Evaporator Type)

A) Feed
B) Vapor
C) Concentrate
D) Steam
E) Condensate
F) Excess Vapor

1) Plate Evaporator
2) Separator

Feed enters at the lower chamber, heats to the boiling point, and then boils as it rises up the space between plates.

2-phase flow of concentrated liquor & vapor exits the top of the plates.

Advantages

- Inexpensive heater design compared to shell & tube
- Compact construction; good for low-head room installations
- Easy to skid-mount
- Can add additional plates for more heating surface
- Sketch shows a rising film; also available in falling-film designs
- Generally HTCs are better than comparable tubular designs

Disadvantages

- High residence times for rising-film
- Limited to concentrates with low suspended solids
- Plate spacing may allow suspended solids to bridge
- Gaskets required between each plate; leaks may develop if the unit will have repetitive heat/cool cycles
- Periodic gasket replacement may be necessary
- HTCs have to be determined by experience or testing

Typical Process Applications

- Food & pharmaceutical , wastewater & chemical applications

Specialty - Thin or Wiped Film Evaporators

Vertical Wiped/Thin-Film Evaporator

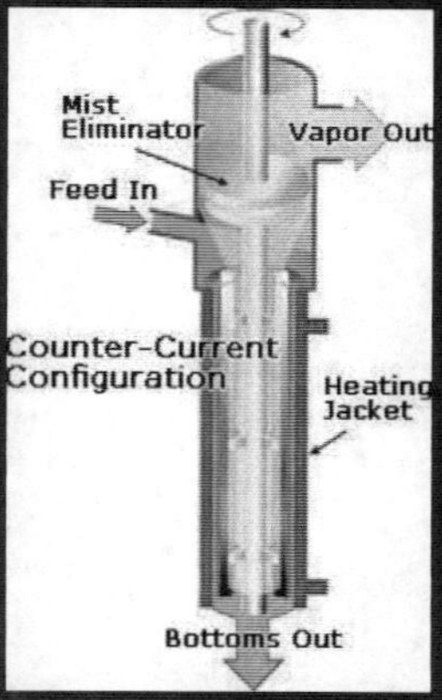

Advantages

- Good for high-solids & scaling services
- Good for viscous liquids (e.g. food, confectionery)
- Short residence time

Disadvantages

- Expensive, high cost per square foot of heating surface
- Horsepower & maintenance intensive
- Scaling or solids buildup requires regular cleaning
  1) Solvent Vapor Out
  2) Process Wall
  3) Rotor Blades
  4) Feed In
  5) Heating Medium In/Out
  6) Bottoms Discharge

Horizontal Wiped-Film Evaporator

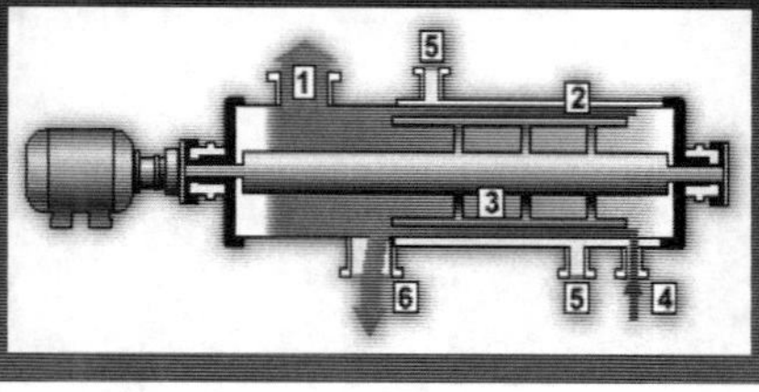

(Sketches by Artisan Industries)

Specialty - Short-Path Evaporator

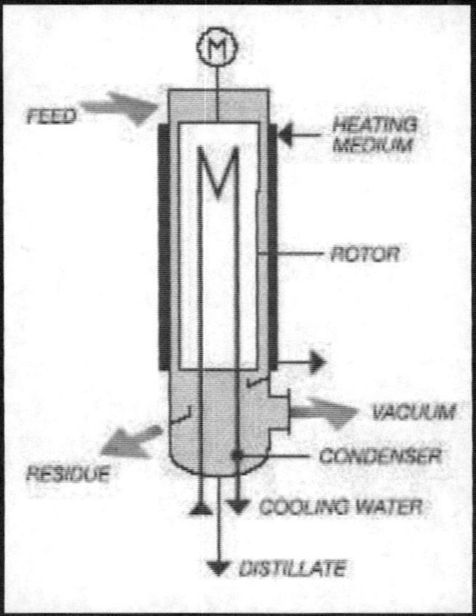

- Short Path Evaporator operation is similar to a thin film evaporator but is used for expensive and/or very heat-sensitive feeds
- Condenser is internal to the evaporator to minimize vapor pressure drop between the condensing surface & the heating surface.
- High operating temperatures & low vacuums (down to 0.001 mbar) required
- Heating medium is usually a heat transfer fluid or high-pressure steam
- Mechanically agitated thin film required to improve the HTC
- Short residence times required to prevent thermal degradation or decomposition
- Short path distillation is used to separate mixtures of organic compounds
- Very expensive cost per square foot of heating surface

(Sketch by InCon Process Systems)

Multiple Effect Evaporators

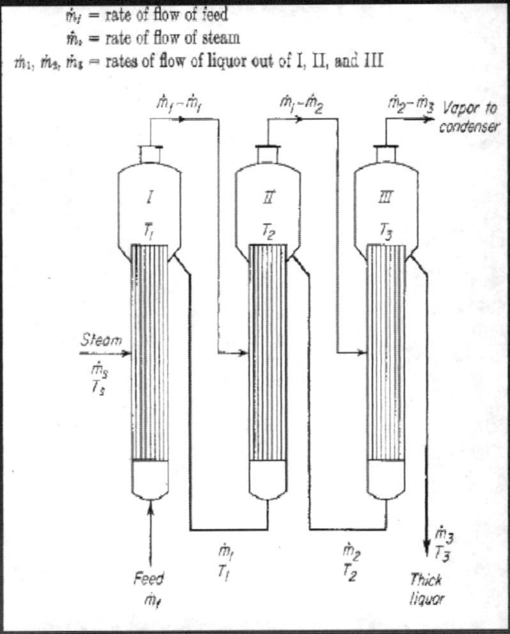

Reference – Unit Operations of Chemical Engineering (McCabe & Smith)

- An effect indicates the presence of a certain vapor pressure for a boiling liquid. The vapor from one evaporator unit may be used to heat & boil a colder liquor at a lower vapor pressure in another unit.
- The 1st effect vapor pressure is higher than the 2nd effect, the 2nd effect is higher than the 3rd, etc.
- The example shown is a 3-effect rising film evaporator. Vapors from the I effect heats the II effect & vapors from the II effect heats the III effect.
- Vapor flow starts with the 1st effect.
- The greater the number of effects for re-using steam, the better the steam economy.
- The last effect vapors will go to a condenser with a vacuum system.
- Rule of thumb – will get approximately 0.7-0.8 lbs of water evaporated per lb of steam used for each effect in an evaporator train. For example, the triple effect shown will evaporate 2.1-2.4 lbs of water for each lb of steam condensed in the first effect heating element.
- The individual effects may be any combination of tubular or plate evaporator types previously shown. Forced circulation types are often used as finishers steps and may be heated separately from the train.

Multiple Effect Evaporator – An American Invention

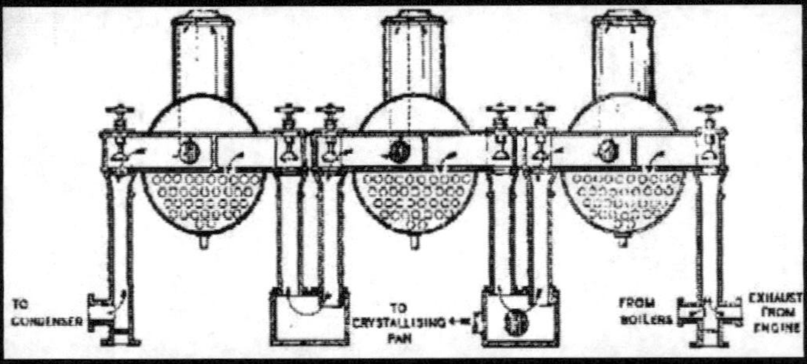

- Norbert Rillieux invented & patented triple effect evaporator which was installed in a Louisiana sugar factory in 1843. The first effect normally took exhaust steam from the engines driving the cane mill. This exhaust could be supplemented when necessary with steam from the boilers. The crystallizing vacuum pans were supplied with vapor off the first body.

Multiple Effect Evaporator – Feed Sequence

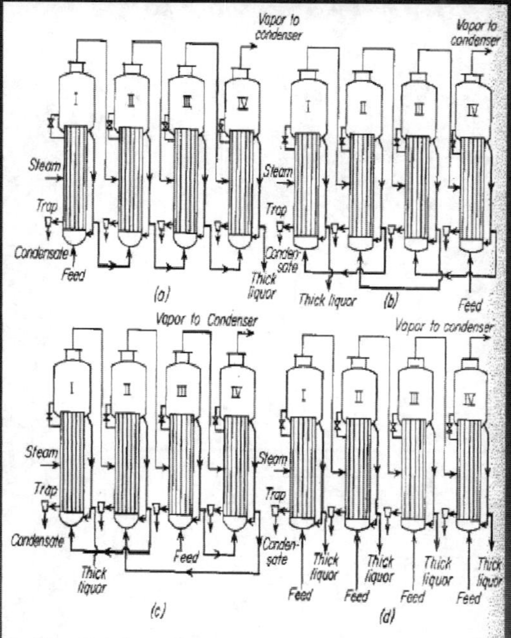

Figure 16-11 *Patterns of liquor flow in multiple-effect evaporators:* (a) *Forward feed.* (b) *Backward feed.* (c) *Mixed feed.* (d) *Parallel feed.*

- Evaporator types are arranged so that the vapor from a previous effect is used to heat the next effect.
- Example shows RF-LTV evaporator types but these may also be FF-LTV, plate type, or FC depending on the process. Some multiple effect trains may have two different evaporator types.
- The feed may be fed in any sequence.
  - Forward feed (a) – liquor transfer I-II-II-IV.
  - Backward feed (b) - liquor transfer IV-III-II-I.
  - Mixed feed (c) - enters a middle effect & go forward or backward to adjacent effects.
- Forward-feed may not require transfer pumps between the effects due to the pressure differences between effects, however, backward or mixed feed will require transfer pumps between the effects depending on the source & end point pressures.

- **Reference – Unit Operations of Chemical Engineering (McCabe & Smith)**

Multiple Effect Evaporator

- The 1st effect has the highest operating temperatures & vapor pressure. The last effect has the lowest temperatures & pressures.
- As the % dissolved solids increases, the liquor boiling point elevation increases across the effects.
- The latent heat in the vapor from each effect increases with decreasing vapor pressures.
- The heat transfer rate to each effect is shown at the bottom of each row
  - For the 1st effect, Q is for the heat from the incoming steam
  - For the subsequent effects, the Q represents the heat from the vapor of the previous effect, flashing liquor feed from the previous effect, & flashing heater condensate from the previous effect
- Evaporator manufacturers using %DS vs HTC information for the liquor will balance the temperature frame & heat transfer rates across the effects so that the heating elements are the same size to simplify the design and minimize fabrication costs.
- Steam economy may be improved by flashing condensate or liquor and re-using the flash steam in a lower pressure effect.

Multiple Effect Evaporator - Temperature Frame

- The spreadsheet example below shows the liquor & vapor temperature distribution for a forward-feed 4-effect evaporator in a corn process. Feed enters the 1st effect & becomes more concentrated by the 4th effect.
- For 40,000 lbs/hr steam rate to the 1st effect, the steam economy would be 3.1 lbs of water evaporated for each lb of steam consumed or 0.77 lbs water evaporated per lb of steam per effect.

Example - 4-Effect Evaporator

Temperature Frame	I	II	III	IV	Total
Steam latent heat, BTU/lb	946.8	969	979.7	997.4	
Steam Pressure, psia	28.8	15.3	10.8	5.7	
Steam Temperature, °F	248	214	197	168	
Temp. Difference,	32	15	26	36	
Liquor Temperature,	216	199	171	131	
Boiling Point Elevation	2	2	3	5	
Vapor Temperature	214	197	168	127	
Vapor Pressure, psia	15.3	10.83	5.72	2.05	
Vapor latent heat, BTU/lb	969	979.7	997.4	1021.6	
Steam rate to steam chest, lbs/hr	40,100	26,453	28,532	32,062	
Feed Rate, lbs/hr	212,000	185,547	157,015	124,953	212,000
Evaporation Rate, lbs/hr	26,453	28,532	32,062	36,620	123,667
Product Rate, lbs/hr	185,547	157,015	124,953	88,333	88,333
% D.S.	20.00	27.00	33.93	48.00	48.00
Heat transfer rate Q, BTU/hr	37,966,680	25,632,957	27,952,800	31,978,639	

Mechanical Vapor Recompression (MVR)

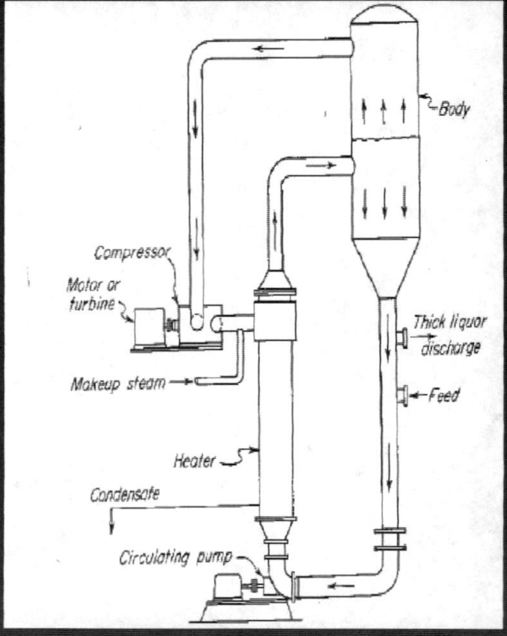

Reference – Unit Operations of Chemical Engineering (McCabe & Smith)

- Rather than using an external steam source, process vapors are compressed & used to heat the process. The motor or turbine is the main energy input.
- Feed is usually pre-heated with hot condensate and/or hot product to minimize the sensible heat loading.
- Vapor is boosted to a higher pressure by a mechanical compressor & the discharge vapor is fed to the heater. The discharge vapor saturation temperature is higher than the boiling point of the liquor for heat transfer.
- Steam addition may be needed for start-up or make-up.
- Depending on the liquor properties, MVRs may be designed for operation at atmospheric pressure or sub-atmospheric pressure if lower operating temperatures are required
- Typical Process Applications
  - Stillage, food, pharmaceutical, distilled water, waste water, pulp & paper, chemical

Mechanical Vapor Recompression (MVR)

- **Advantages**
  - Energy utilization is very good. Depending on the feed & evaporator configuration, the economy can correspond to 10-20 effect evaporator.
  - Cost effective where $/kW-hr cost is low
  - Tubular & plate falling-film evaporators are commonly used for MVR applications since they can operate with lower ΔT's & higher HTCs. FCs are also used.
  - Condenser not required & cooling water required is minimized.
  - Vacuum pump not required for atmospheric operation
- **Disadvantages**
  - High initial cost.
  - Sub-atmospheric MVRs require larger compressors, vapor bodies, & vapor piping.
  - Less economical for high-boiling point liquors
  - Compressor requires additional controls & maintenance
  - Inlet vapor to compressor may require superheating to eliminate entrained droplets. Discharge vapor may require de-superheating to avoid dry tube surfaces.
  - High voltage motors may be required for large motor sizes

Questions

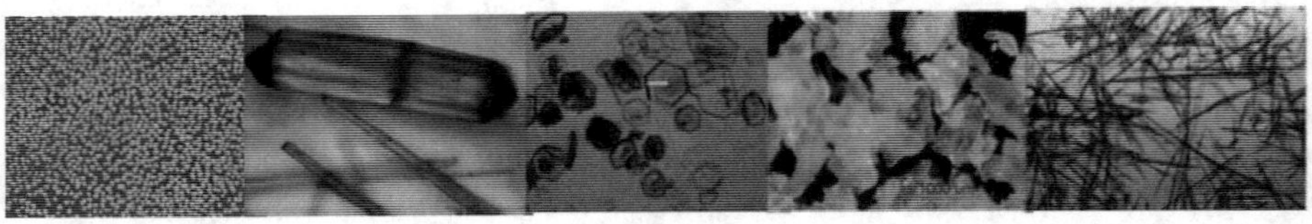

Tips and Tools for Solution Crystallization Process Design

Paul Larsen, Jaime Curtis-Fisk, Megan Donaldson, Timothy Frank, Jörg Heinrich, Zubin Kuvadia, Ron Leng, Patrick McGough

AIChE Process Development Symposium; Chicago, IL; June 12, 2013

Why Consider Solution Crystallization?

- Product is a crystalline solid or slurry
  - Pharmaceuticals
  - Agricultural actives
  - Electronic materials
  - Catalysts

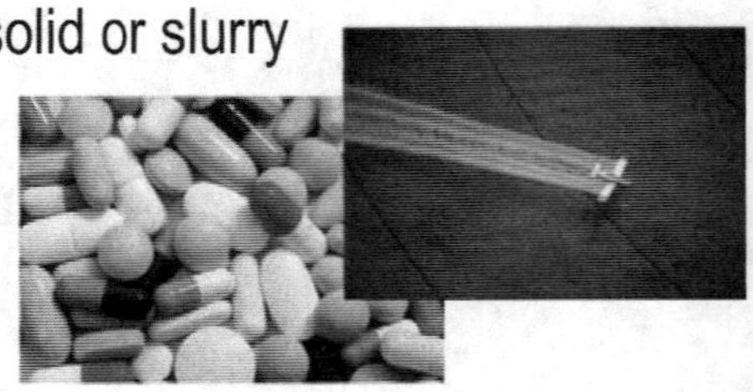

- Distillation not attractive
  - Relative volatility of key components < 1.3
  - Product is temperature-sensitive (decomposition, reaction)
  - Components have low volatility
- Potential for high degree of purification in single processing step
- Lower operating temperature than melt crystallization

Crystallization Processes: The Big Picture

- The best solvent for the reaction may be bad for crystallization
- Downstream unit ops often have significant impact on product properties
- A long crystallization cycle time may result in shorter overall cycle time
- ***Overall process should be optimized***

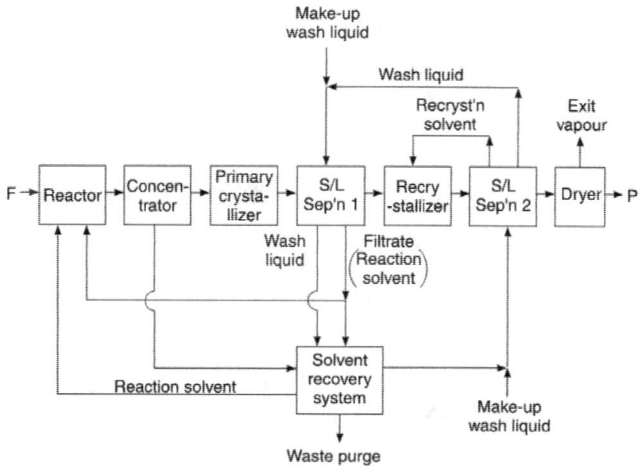

Figure Source: W. Chang and K. Ng, *AIChE Journal*, Vol 44, No. 10, 1998

Crystallization Fundamentals

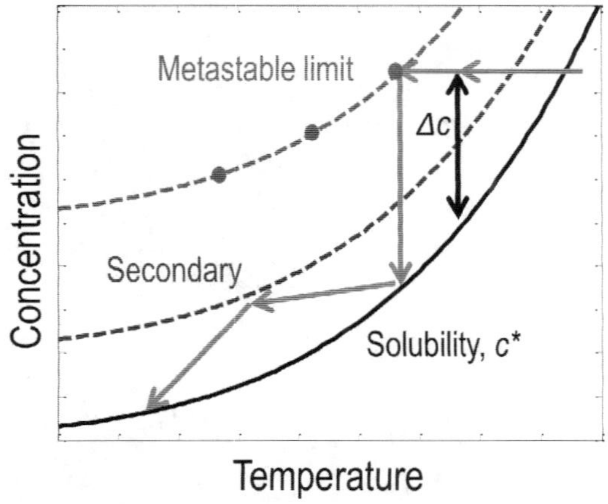

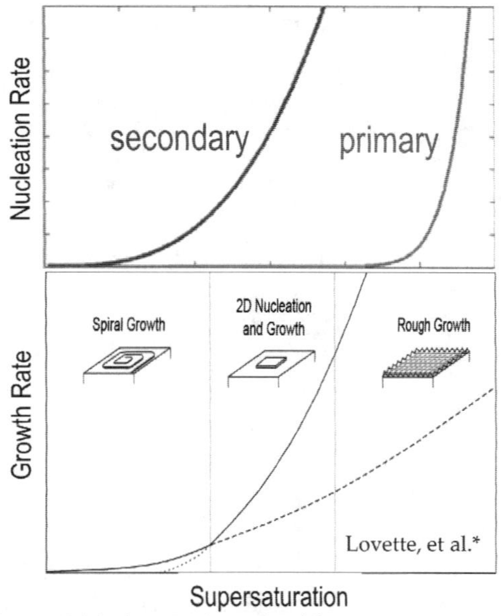

*Adapted with permission from M.A. Lovette, A.R. Browning, D.W. Griffin, J.P. Sizemore, R.C. Snyder, M.F. Doherty, "Crystal Shape Engineering," *I&ECR*, 47, 9812-9833. Copyright 2008 American Chemical Society.

Crystallization Process Design Workflow

Ideal Process Design

In reality, the process is often designed iteratively

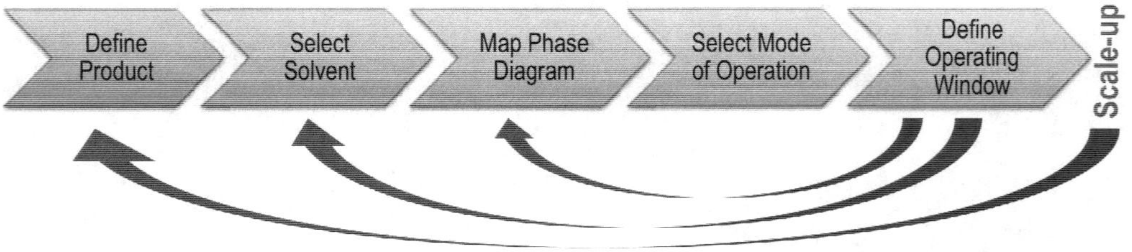

Effective execution of each workflow step will minimize costly recycle

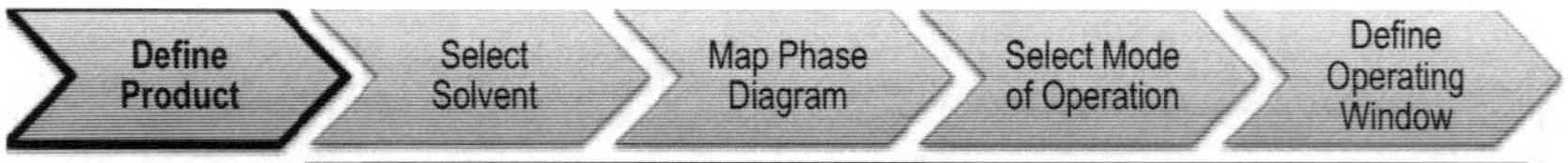

Motivation: Solid form (crystal structure, size, shape, composition) impacts product performance; ease of processing and formulation; IP

Graphite →

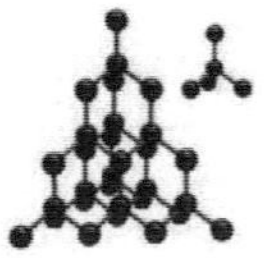

Diamond →

Tips: Initiate collaboration *early* with Analytical and Formulations to understand stability and performance of solid forms

Tools: High throughput screening; PXRD; TGA; DSC; Raman; Accelrys Materials Studio

Example: Polymorph identification for commercial pesticide

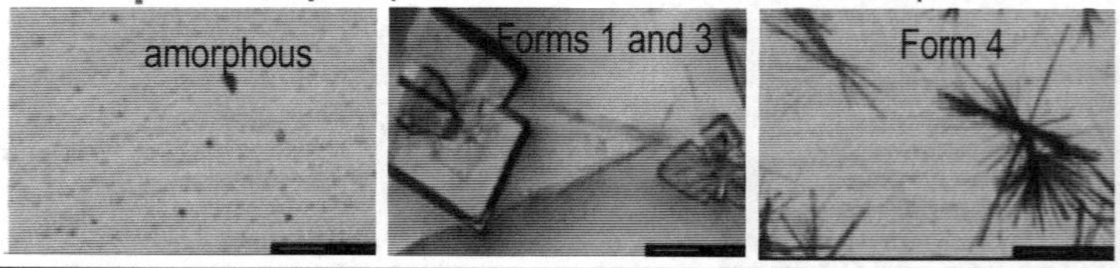

Acetic Acid | THF | Dichloromethane

Define Product > **Select Solvent** > Map Phase Diagram > Select Mode of Operation > Define Operating Window

Motivation: Solvent impacts solubility (yield), growth rate (capacity), nucleation rate (PSD), crystal structure and shape (product properties)

Tips:

- Engage *early* with chemist to influence solvent selection
  - Simple overall process scheme
  - Same solvent for reaction and isolation is desired
- Consider solvent recovery and EH&S properties
- Evaluate impact of solvent on particle shape

Tools:

- Solvent selection guides / databases (e.g. ACS GCIPR; ICH Guidelines)
- Solubility prediction tools to explore solvent combos
  (N. Chennamsetty et al., AIChE Annual Meeting 2011; K. Joback, AIChE PDS 2012)
- Heuristics/short-cut methods based on crystal structure
  (P. Dandekar et al., *Materials Research, 2013*)

Example: Effect of co-solvent on crystallization of active ingredient

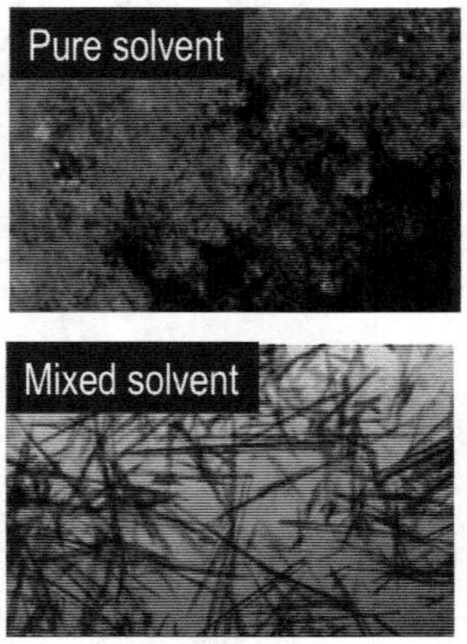

Define Product > Select Solvent > **Map Phase Diagram** > Select Mode of Operation > Define Operating Window

Motivation: The phase diagram determines maximum theoretical yield, dictates mode of operation, and defines the window of operation

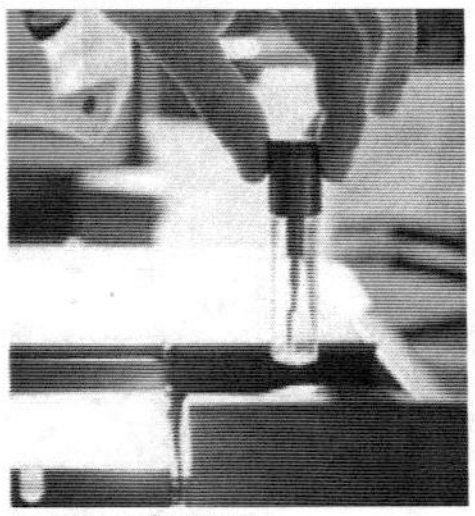

Avantium Crystalline

Tips:

- Invest time up-front to characterize the phase diagram in the region of interest
- Consider impact of impurities
- Confirm solid is desired product
- MUST validate predicted solubilities experimentally

Tools:

- Analytical: ATR-FTIR, Turbidity, DSC
- Modeling: OLI Analyzer Studio, SLEEK, CRANIUM
- Experimental: Avantium Crystalline, HEL CrystalSCAN
- Experimental design methodologies (P. Barrett et al. *Chem Eng. Res.& Des.* 2002; A.R. Parsons et al., *Chem Eng. Res. & Des.* 2003; Y Yi et al., *IECRES* 2005; M. Rabesiaka et al., *J. Crystal Growth* 2011)

Example: Effect of impurities on solubility

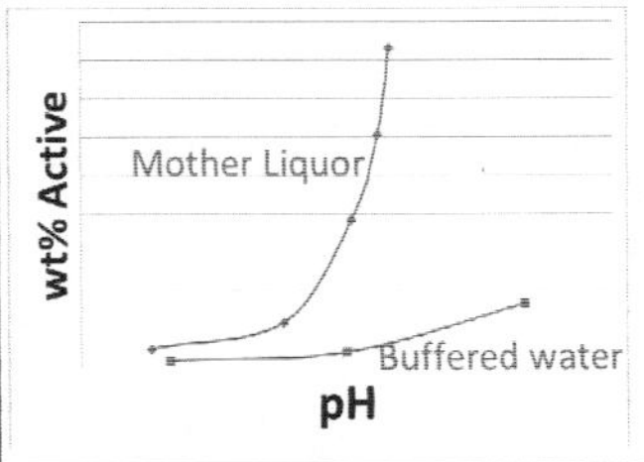

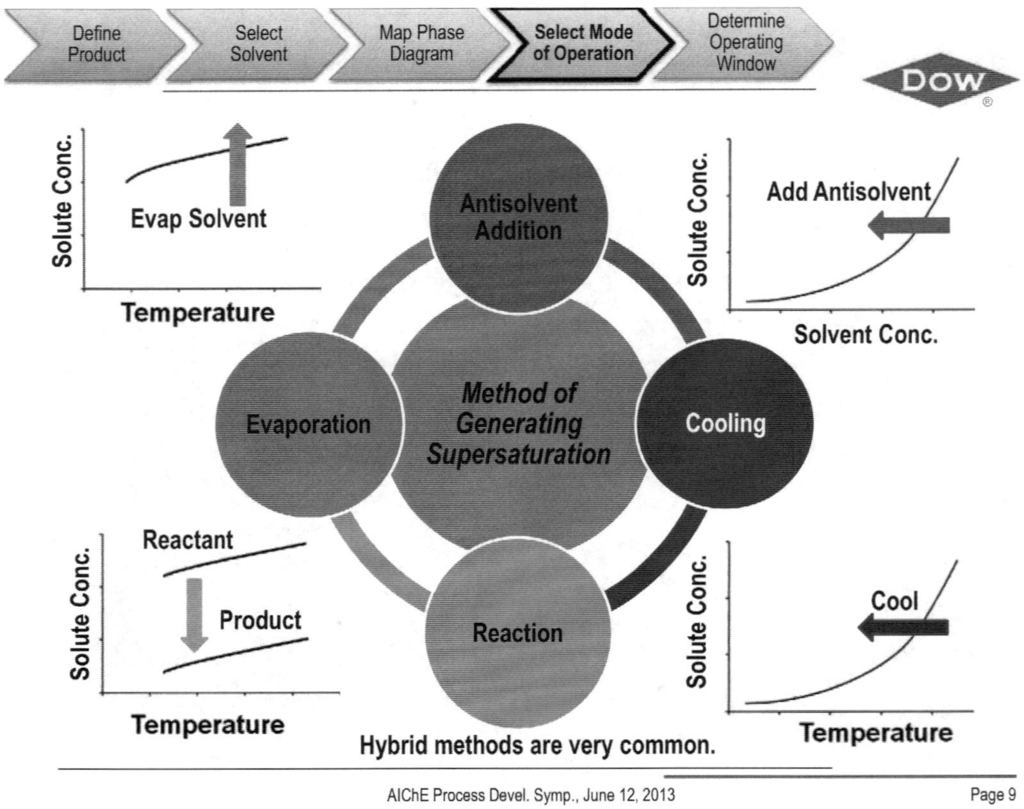
Define Product
Select Solvent
Map Phase Diagram
Select Mode of Operation
Determine Operating Window
DOW
Solute Conc.
Evap Solvent
Temperature
Antisolvent Addition
Solute Conc.
Add Antisolvent
Solvent Conc.
Evaporation
Method of Generating Supersaturation
Cooling
Solute Conc.
Reactant
Product
Temperature
Reaction
Solute Conc.
Cool
Temperature
Hybrid methods are very common.
AIChE Process Devel. Symp., June 12, 2013
Page 9

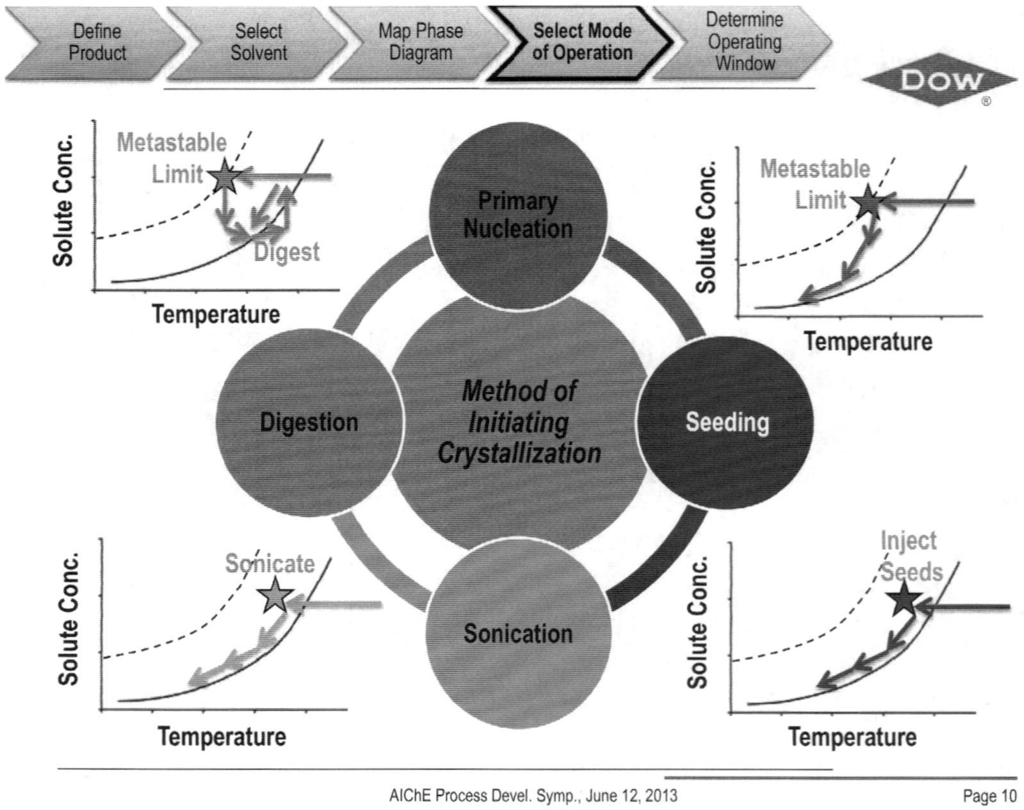
Define Product
Select Solvent
Map Phase Diagram
Select Mode of Operation
Determine Operating Window
DOW
Solute Conc.
Metastable Limit
Digest
Temperature
Primary Nucleation
Solute Conc.
Metastable Limit
Temperature
Digestion
Method of Initiating Crystallization
Seeding
Solute Conc.
Sonicate
Temperature
Sonication
Solute Conc.
Inject Seeds
Temperature
AIChE Process Devel. Symp., June 12, 2013
Page 10

Define Product > Select Solvent > Map Phase Diagram > **Select Mode of Operation** > Determine Operating Window

Motivation: Method of generating supersaturation impacts yield, operability, and ease of scale-up. Method of initiation impacts process variability, product properties.

Tips:

- Select cooling crystallization when possible
- Avoid primary nucleation-based initiation due to high variability
- High seed loadings (20-50% of total solids) may be necessary for systems with very low solubility
- Test various modes and configurations experimentally
- Experimental studies should simulate production environment

Tools:

- Glass, thermostatted, 50 mL—1 L reactor systems (Mettler-Toledo EasyMax, OptiMax; Syrris Atlas; HEL AutoLAB);

Define Product > Select Solvent > Map Phase Diagram > **Select Mode of Operation** > Determine Operating Window

Continuous Crystallization

- Large volume, commodities
- High productivity
- Supersaturation via evaporation (typically)
- Dedicated to single product
- Encrustation/plugging issues
- Overall flowsheet is continuous

Batch Crystallization

- Small volume, specialties
- Low productivity
- Supersaturation via different methods
- Flexible, multi purpose
- Dissolve encrustation in between batches
- Overall flowsheet is batch
- Contract manufacturing

Define Product → Select Solvent → Map Phase Diagram → **Select Mode of Operation** → Determine Operating Window

DOW®

Non-traditional Crystallizers

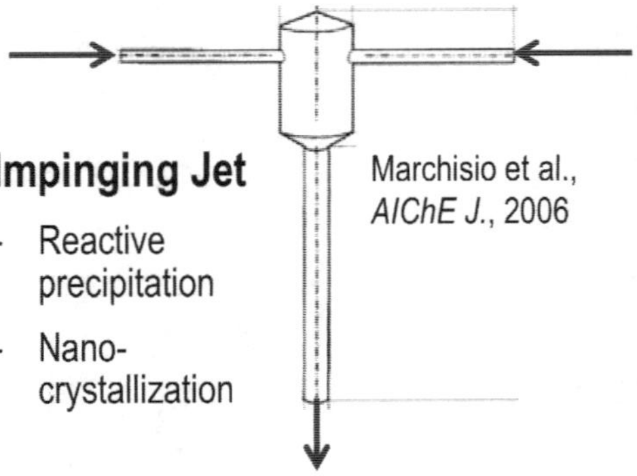

Impinging Jet

- Reactive precipitation
- Nano-crystallization

Marchisio et al., *AIChE J.*, 2006

Oscillatory Baffle Tube Crystallizers

- NiTech, Cambridge Reactor Design
- Compact, reduced residence time
- Continuous operation

Define Product → Select Solvent → Map Phase Diagram → Select Mode of Operation → **Determine Operating Window**

DOW®

Motivation: The operating rates and mixing environment determine the kinetics of crystal nucleation, growth, agglomeration, and attrition, ultimately impacting crystal purity, size, shape, and structure.

Tips:

- Rate of crystallization is directly proportional to available surface area ➔ need to operate at lower rates in initial stages of crystallization; consider seeding
- For batch processes, use non-linear addition/cooling profiles based on curvature of solubility curve (E. Aamir, C.D. Rielly, Z.K. Nagy, *IECRES* 2012 51(51) 16677-16687)
- Point of addition matters

Tools:

- Glass, thermostatted reactor systems; 1-L scale
- In situ particle characterization: Laser back-scattering (e.g. FBRM), video imaging, turbidity, Raman
- Solution concentration monitoring: ATR-FTIR, Raman, sampling
- Experimental design methods (Teich et al., AIChE Annual Meeting 2010)

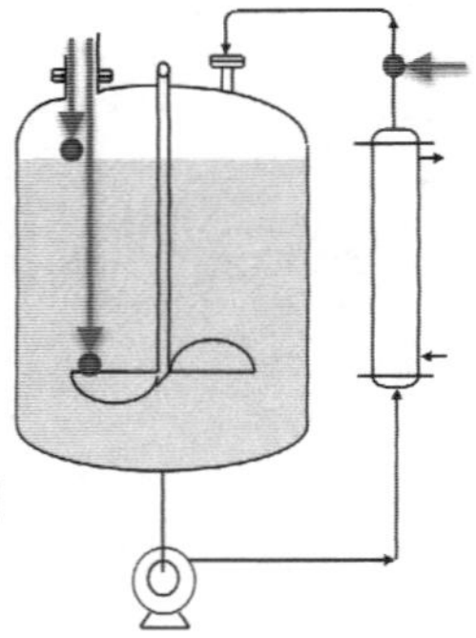

Define Product > Select Solvent > Map Phase Diagram > Select Mode of Operation > **Determine Operating Window**

Example: Crystallizer Capacity Improvement

- Batch Crystallization
- 2,000 gallon, jacketed CSTR
- Supersaturation generated by reaction (pH swing) and cooling
- Crystallization initiated by primary nucleation
- Variability in feed and product
- Capacity increase necessary to meet growing demand

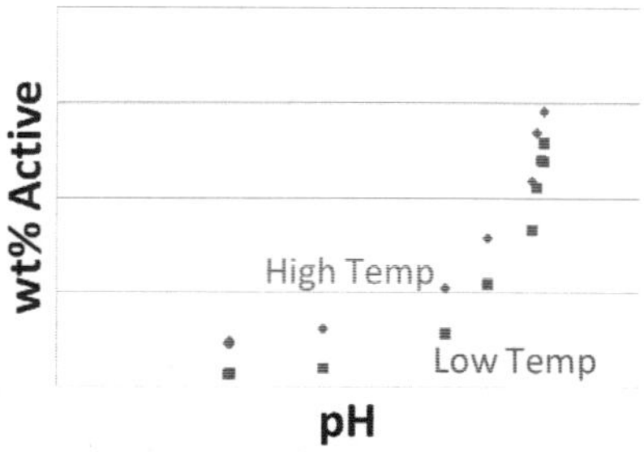

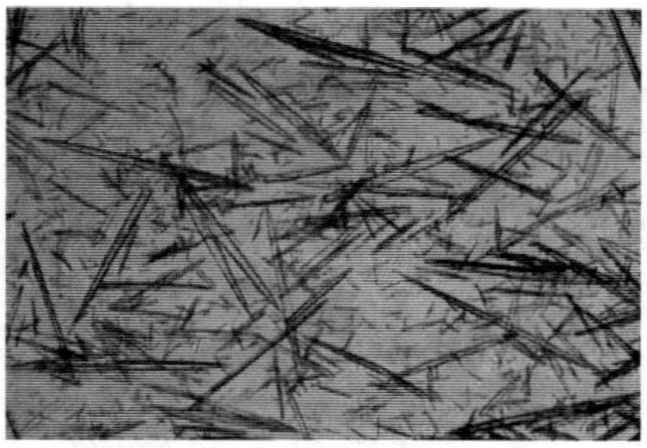

Define Product > Select Solvent > Map Phase Diagram > Select Mode of Operation > **Determine Operating Window**

Example: Crystallizer Capacity Improvement

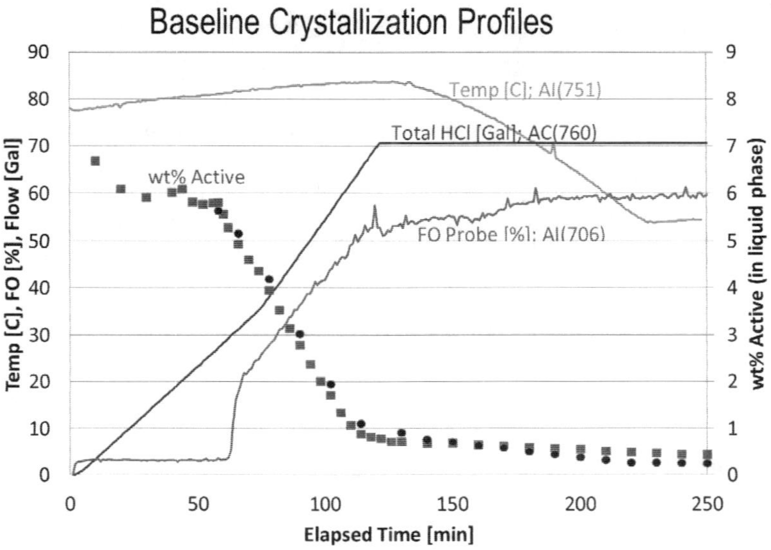

Define Product → Select Solvent → Map Phase Diagram → Select Mode of Operation → **Determine Operating Window**

DOW®

Example: Crystallizer Capacity Improvement

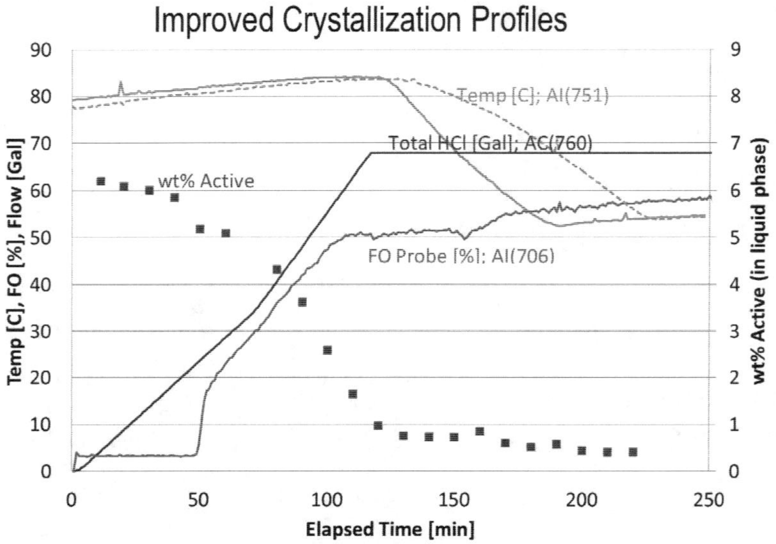

Results:

- 10% reduction in cycle time
- Negligible impact on filtration and yield

DOW®

Summary

Define Product → Select Solvent → Map Phase Diagram → Select Mode of Operation → Define Operating Window → Scale-up

- Consider impact of process decisions on crystallization to achieve an optimum overall process
- Yield, operability, purity, and cycle time are often just as important as particle size.
- Invest time to understand the phase diagram
- Test various modes of operation experimentally
- Take advantage of advanced experimental and modeling tools

Crystallization Literature Resources

- Crystallization Text Books:
  - Jones, A. G.: *Crystallization Process Systems*, Butterworth-Heinemann Oxford, 2002
  - Mersmann, A.: *Crystallization Technology Handbook*, 2nd Edition, Marcel Dekker New York, 2001
  - Mullin, J. W.: *Crystallization*, 4th Edition, Butterworth-Heinemann Oxford, 2001
  - Myerson, A. S. (ed.): *Handbook of Industrial Crystallization*, 2nd Edition, Butterworth-Heinemann Boston, 2002
  - Söhnel, O., Garside, J.: Precipitation - Basic Principles and Industrial Application, Butterworth Heinemann Oxford, 1992
- Encyclopedias:
  - Mullin, J. W.: Crystallization and Precipitation, in: Ullmann's Encyclopedia of Industrial, Chemistry, 7th Edition, Wiley Weinheim, 2007
  - Ulrich, J, Stelzer, T.: Crystallization, in: Kirk-Othmer Encyclopedia of Chemical Technology, Volume 8, Wiley Weinheim, 2011
  - Perry, R. H., Green, D. W. (eds.): Perry's Chemical Engineers' Handbook, 8th Edition, McGraw-Hill New York, 2008
- Crystallization Papers: see www.sciencedirect.com

Innovations in Mixing for Unmet Needs of Manufacturing and R&D

Kishore Kar

Fellow

The Dow Chemical Company

Presentation Outline

- Industrial Perspective of Challenges in Mixing
- Dow Fluid Mechanics and Mixing R&D - Innovations
- KT-3: Tickler Impeller for Minimization of Solids Heel
- KDM: Mixing Device for High Throughput Research (HTR)

Industrial Perspective of Challenges in Mixing

A. Technical Challenges

- Scale-up / Scale-down
  - Understanding fundamental parameters for scaling from a flask
  - Transfer of knowledge
- Multi-phase Mixing in Reactors and Crystallizers
  - Reactive mixing liquid-gas-solid systems
- Liquid-Liquid Dispersion Using High Shear Dispersers
  - Drop size distribution (DSD) kinetics
  - High O/W concentration phase ratio
- CFD: Limitations of RANS Models for STR
  - Computational resources for LES and DNS are a bottleneck

64 Hu

6/20/2013 3

B. Existing Asset Utilization and Capital Constraints

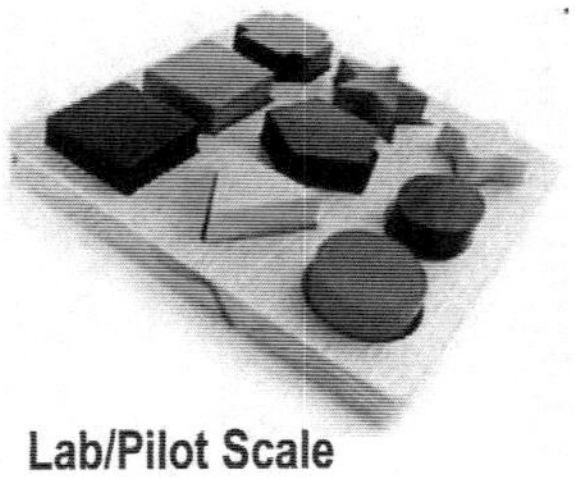

Lab/Pilot Scale
Looks good-success

Production Scale
- **Assets are in place**
- **No or little capital**

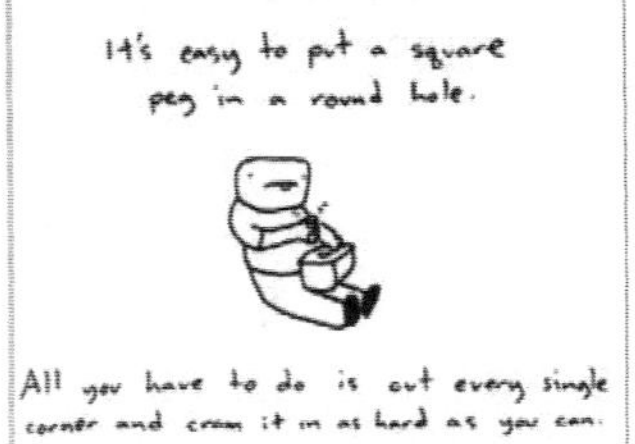

64 Hu

6/20/2013 4

Industrial Examples

Approx. 50 vol.% of Dow products are solids!

Common Multiphase Mixing Processes

Large Scale Processes:

- Terephthalic Acid (CTA) for Polyester*
- High Density Polyethylene (HDPE) – Slurry-Phase process
- Polystyrene ➔ Styrofoam®

***Global PTA Production: 110 billion pounds/yr.**

Polyethylene Terphthalate (PET) Process

Process Steps:

Crude terephthalic acid → (hydrogenation) Purified terephthalic acid → (monoethylene glycol) Polyester

CTA Process: Liquid – Gas – Solid Multiphase Reaction

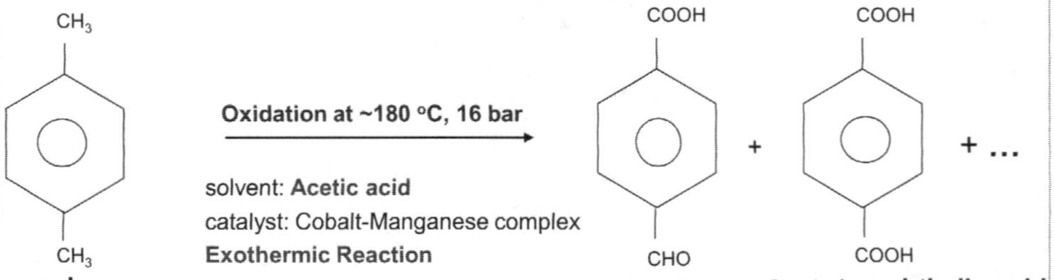

***p*-xylene**

4-Carboxy-benzaldehyde (4-CBA impurity)

Crude terephthalic acid

KT-3: A Novel Tickler for Slurry Vessels

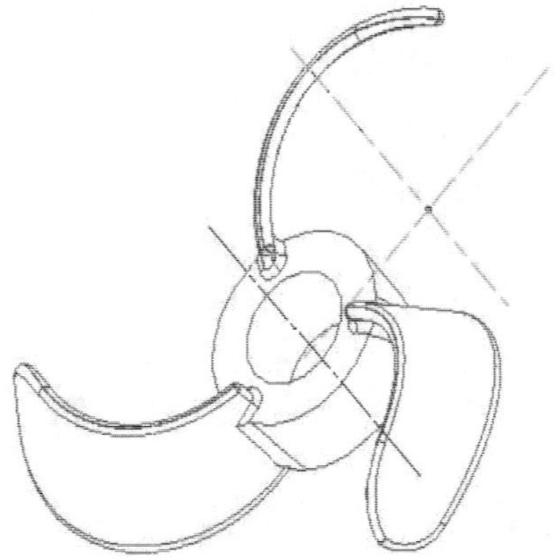

Innovations for manufacturing application

Tickler impeller and its function?

- smaller agitator than the main impeller(s)
- located close to the tank bottom
- suspends solids when slurry level is below the lower main impeller
- mixes even the last remains of slurry

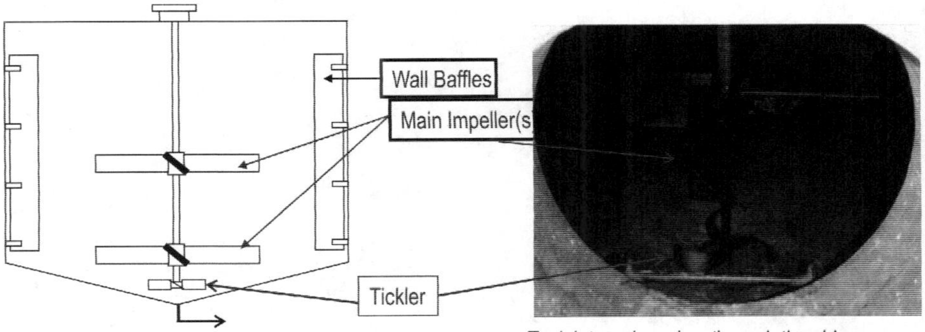

Tank internals – view through the side manway

Conventional Ticklers and Newly-Developed KT-3 Tickler

Pitch Blade Turbine (PBT)

Flat Blade Turbine (FBT)

Limitations of Radial Discharge
solids thrown away from exit nozzle
low liquid level breaks pump suction

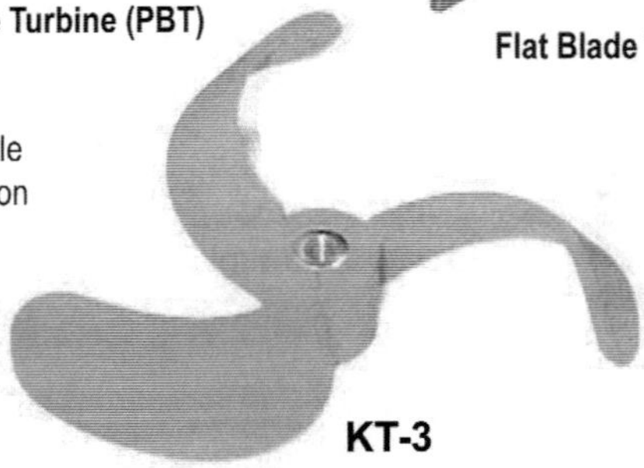

KT-3

6/20/2013 9

What Spawned KT-3 Development?

Solids Heel ➔ Process Solids Loss: ~1 MM lbs/yr

Slurry: 30-40 wt% solids in water

Solids: ρ = >1.3 g/cc; d_{32}= 260 mm

Vessels diameters: ~120 – 216 in.

6/20/2013 10

Scale-Down Experimental Studies

- **20,000 gal ➔**
  - **100-gal (30" ID) lab scale**
  - **500-gal (60" ID) pilot plant**
  - **Understanding the defects**
  - **Conceptualization of a new Tickler impeller**
  - **Performance validation**

100-gal Lab scale tank

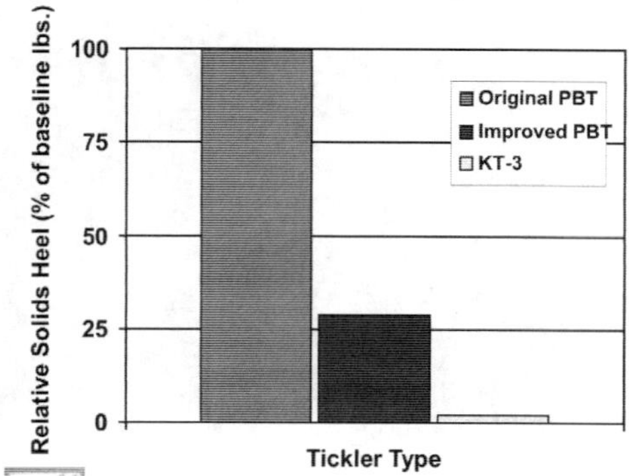

Implementation in the concerned plant and leveraged …

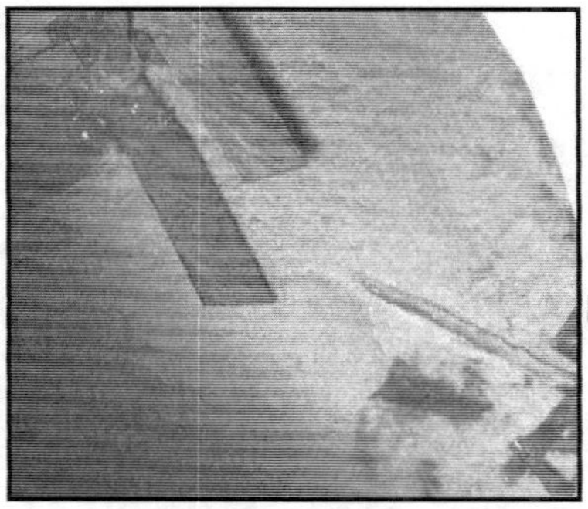

before KT-3 installed: solids heel >1000 lbs

after KT-3 installed: solids heel <5 lbs

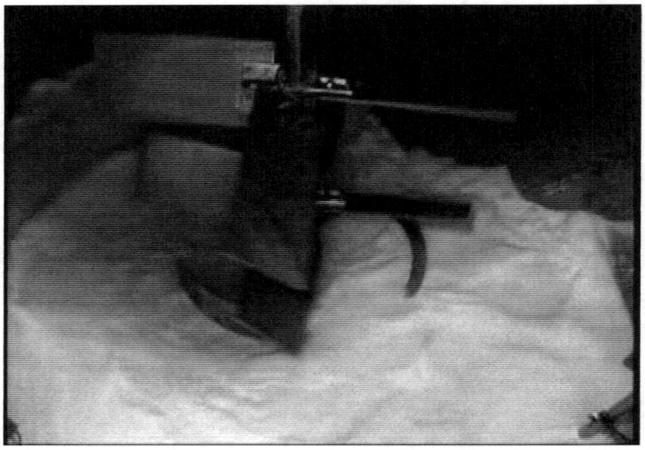

Movie showing sweeping flow pattern (works in slide mode)

6/20/2013 13

Various KT-3 Ticklers in Use

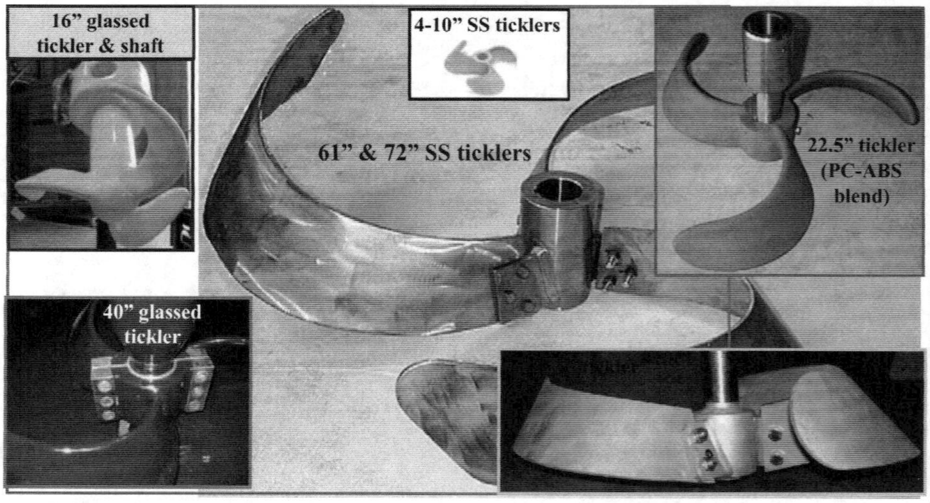

Turbulent Flow Power number = 0.6

Alternate Application:
KT-3 Tickler Assistance in Draining Sticky, High-Viscosity Fluid

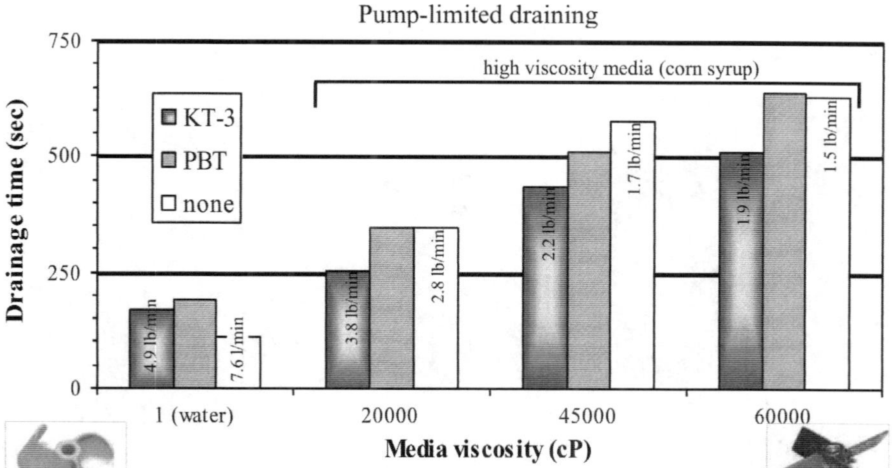

KT-3 tickler

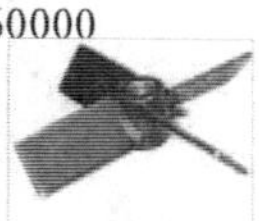
PBT tickler

T=18"; D/T=0.33; N=125 RPM; N_{Re}=152000, or ≤10

KT-3 Summary

- Patented technology (US Patent 6,955,461 B2) that:
- is plant and vendor tested and proven
- minimizes/eliminates solids heels
- provides more stable slurry feed rate during draining
- dramatically reduces pump starvation due to vortexing
- enhances pump-limited draining of sticky, high-viscosity fluids

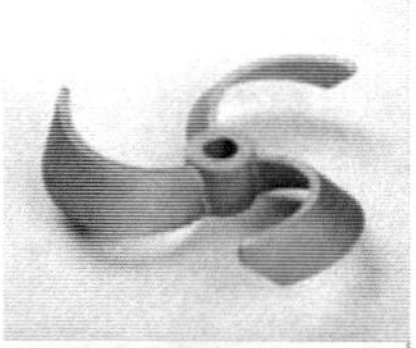

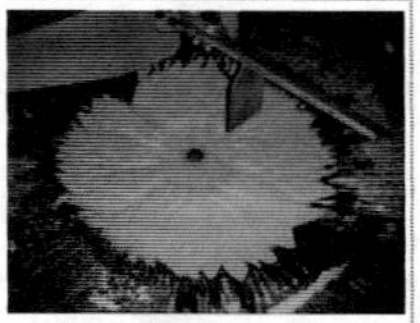

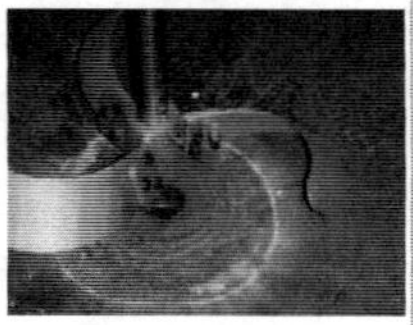

KDM: A Novel Impeller for High Throughput Research

HTR - Parallel Pressure Reactor (PPR)

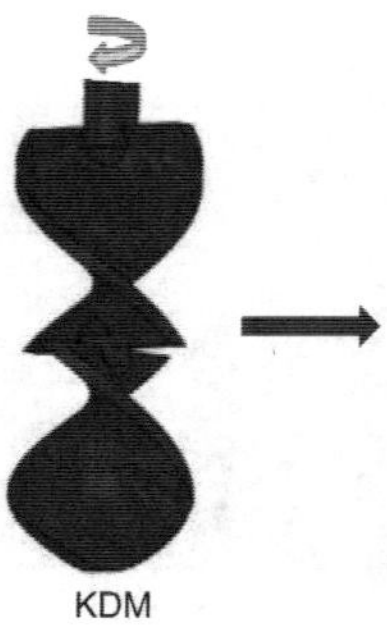

KDM

Innovations for R&D application

*** HTR – High Throughput Research**

What Spawned KDM Development?

- Lack of efficient agitation system for unbaffled vials used in the HTR devices using high viscosity media
- Mitigate mixing controlled phenomena such as mass transfer and dissolution
  - Catalyst reaction kinetics rate Studies
  - Formulation – dispersion and dissolution efficacy

What is the KDM Mixer?

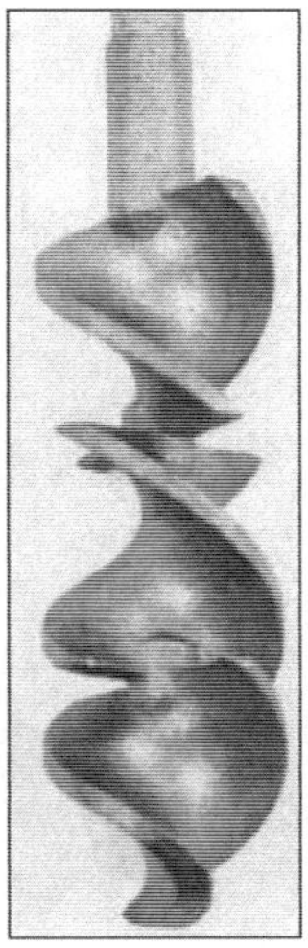

- **Shaft-mounted twisted-ribbon impeller**
- **Not a helical ribbon or auger**
  - ***not a blade winding around a shaft***
  - ***critically, successive elements off-set by 90°***
- **Not just a static mixer on a shaft**
  - ***critically, all elements pump in same direction***

Experimental Set-up and the Commercial HTR Mixers

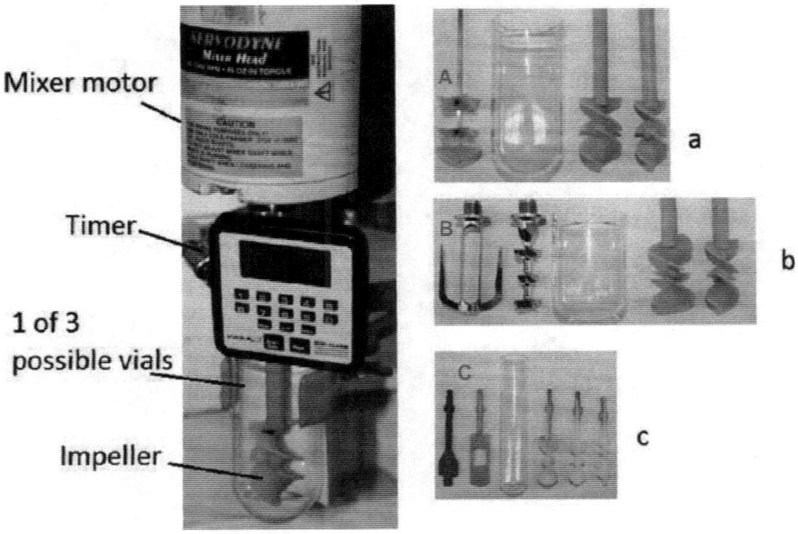

Comparison of Mixing Performance

0 seconds **2 seconds** **5 seconds**

Dow (in house)
1" KDM
15180 cP viscosity
1.41 N-cm torque
0.30 W power

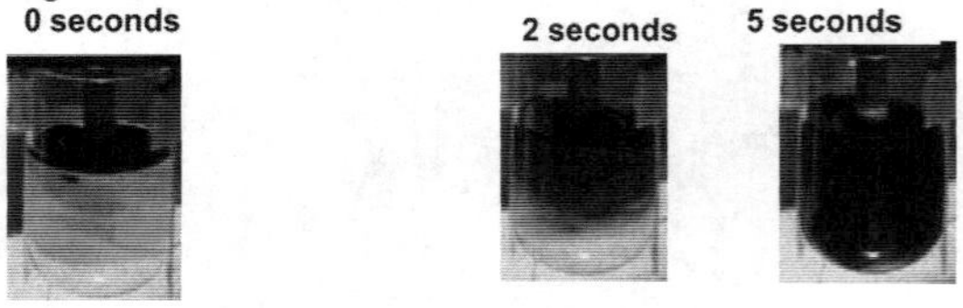

Vendor SysA
9/8" PBTs/paddle
15180 cP viscosity
1.41 N-cm torque
0.30 W power

Vendor SysB
¾", ⅝", ¾" PBTs
15460 cP viscosity
0.35 N-cm torque
0.07 W power

Vendor SysB
1.5" Anchor
15460 cP viscosity
3.71 N-cm torque
0.78 W power

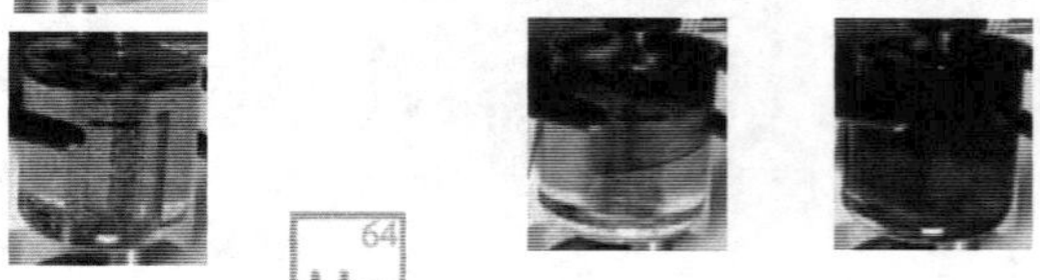

Effects of KDM Parameters on Mixing Performance

- Diameter ratio (D/T)
- Liquid submergence (s)
- Off-bottom clearance (c)
- Length-diameter ratio (L/D)
- Number of elements

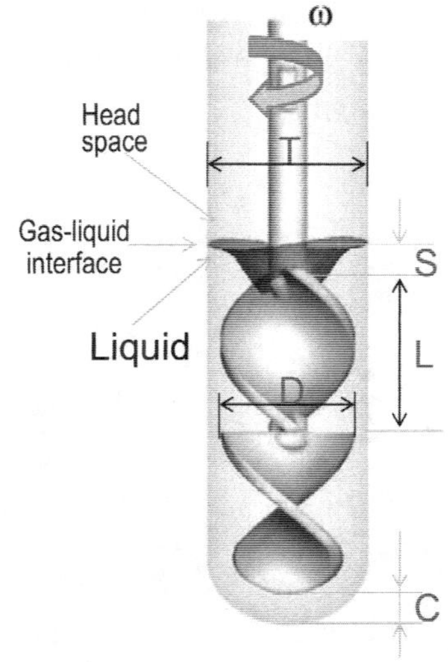

Comparing CFD Predictions with the Experiments for 3-Element KDM

D=0.33", D/T = 0.55, L/D = 1.2

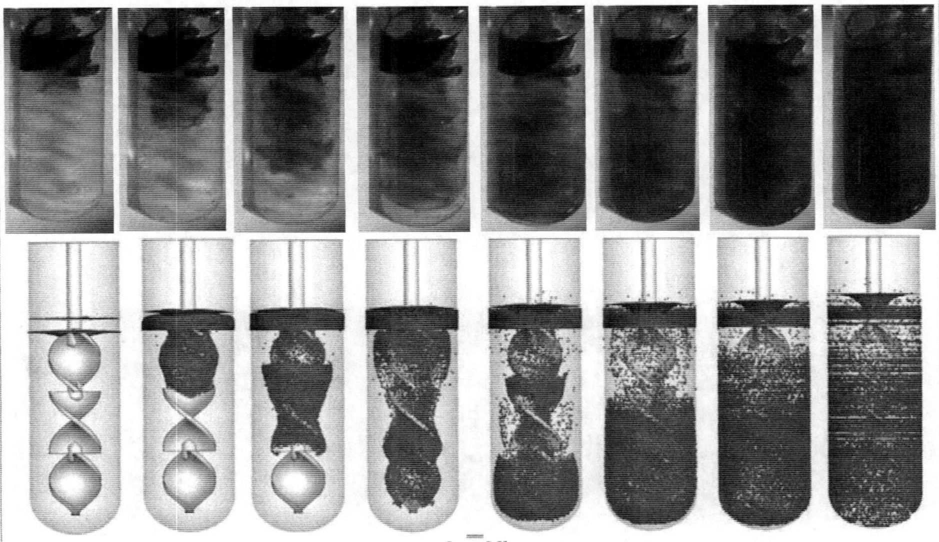

0s 1s 2s 3s 5s 7s 10s 20s

Effect of Impeller/Vessel Diameter Ratio (D/T)

Despite down-pumping operation, flow is upward at centerline w/ choking (*D/T ~ 0.6 ~ optimal*)

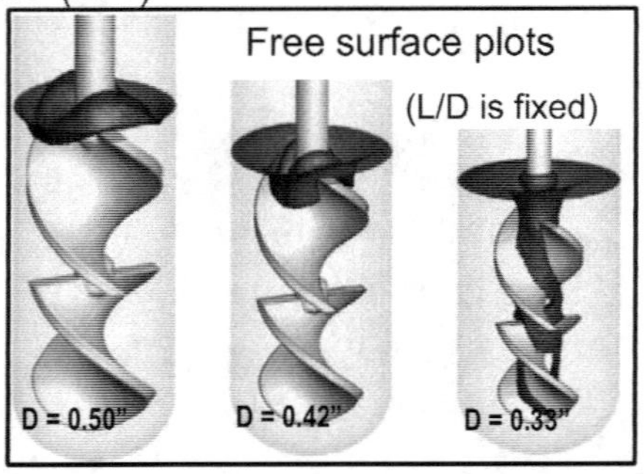

Axial velocity contours on centerline plane

(L/D is fixed)

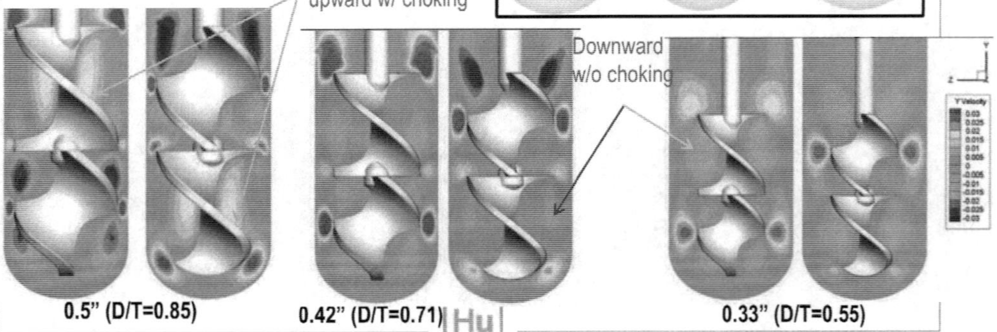

Effect of Pitch Angle: Optimizing Element Length (L) and Diameter (D)

0.6" vial, 30,000cp, 400rpm

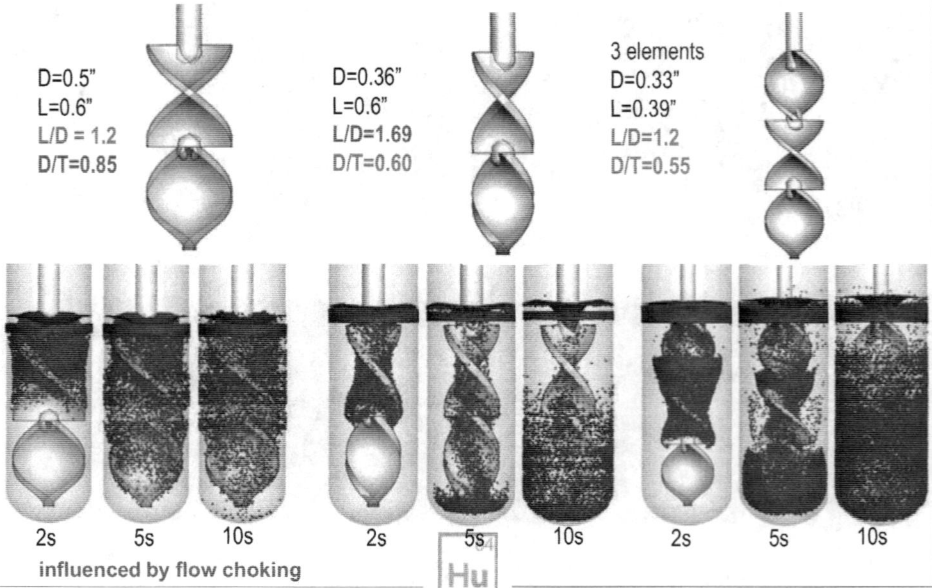

KDM Characterization: Power Number Vs. Reynolds Number (Laminar Zone)

$$N_P = \frac{P}{\rho N^3 D^5}$$

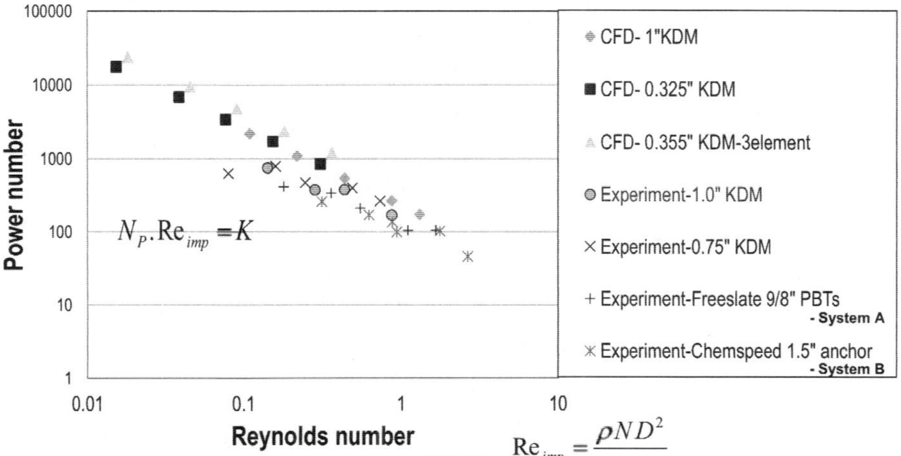

$$\mathrm{Re}_{imp} = \frac{\rho N D^2}{\mu}$$

KDM Implementation in HTR Device

KDM Summary

- Patented (US 7, 946, 753 B2) technology that:
- is lab tested and proven as HTR mixer
- superior mixing, solids dispersion and gas drawdown performance
  - → minimizes risk of reaction kinetics to be mass transfer controlled
  - → eliminates the risk of dissolution to be dispersion controlled

6/20/2013 29

Acknowledgements

Dr. Richard Cope
Dr. Zhao Yu
Dr. Ralph Evens

E&PS - Fluid Mechanics & Mixing Discipline Members at Dow

Ref. KT-3: A Novel Tickler for Solids Removal from Slurry Vessels, **Ind. Eng. Chem. Res.** ,2009, 48, 4990-4997
Ref. Mixing Performance of the Kar Dynamic Mixer Impeller in Small Laboratory-Scale Systems, , **Ind. Eng. Chem. Res.** ,2012, 51, 15282-15292

6/20/2013 30

Basics of Twin Screw Extrusion Compounding

Kalman, D., Solvay

Originally presented on: 6/12/2013 16:30:00 - 17:00:00

Twin screw extrusion compounding is used in a wide variety of processing applications, from compounded polymers for injection molding and other end uses, to polymer degassing and solvent removal, pharmaceutical hot melt extrusion, and even cereal and pet food production. This talk will focus on twin screw extrusion for compounding engineering polymers. The overall layout of a compounding production line will be presented with discussion of issues in the overall process, from feeding to extrusion to pelletizing. The key design parameters for twin screw extrusion are presented, including geometry, free volume, various screw elements, screw design, die design, pelletizing, and how to define the goals of your compounding process and select the appropriate design details. In addition, the key process parameters will be laid out and their effect on each other will be discussed. Methodology for scale-up from R&D to manufacturing will be presented. The differences between twin screw and single screw extrusion will be discussed with information on other extrusion processes.

SOLVAY
asking more from chemistry®

Overcoming Agitation Challenges

Cold Pearl Surfactant Blends

20/06/2013

Ian Lutz

Pearlizers

- Additives of some kind used to impart a shiny, iridescent quality to a wide range of products (paints, plastics, cosmetics/home/personal care)

- Multiple reflections on semi-transparent surfaces in suspension

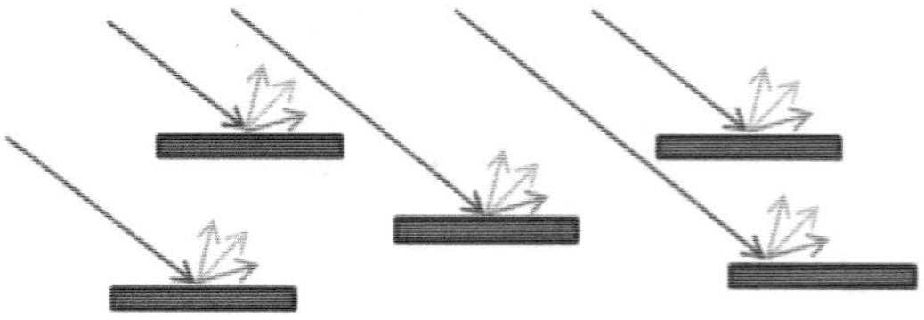

2

SOLVAY
asking more from chemistry®

Pearlizers in personal care, a brief history*

- Powdered bismuth was first used in Egypt to add pearlescent quality to their wide range of cosmetics
- Jean Passieu, a French fisherman, isolated crystals from fish scales for use as a pearlizer in cosmetics
  - 100 tonnes of fish for 2 kg of pearlizer
  - Smelled of fish if not properly extracted

- Synthetic means were pursued as a result
  - Zinc, lead, mercury, mica were originally used for cosmetics
  - Crystallized magnesium stearate, alkanolamide, **ethylene glycol distearate** for shampoos

3 | *CROMBIE, R. L. Cold Pearl Surfactant Based Blend International Journal of Cosmetic Science 1997, 19, 205-214.

SOLVAY
asking more from chemistry®

Cold Pearl Overview

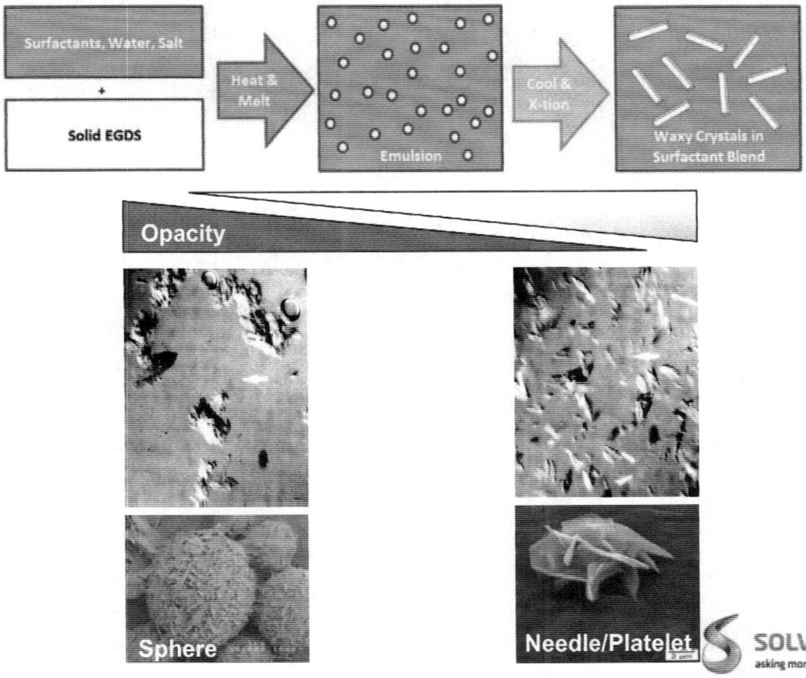

4 |

SOLVAY
asking more from chemistry®

Grading Pearlizers

- 0-5 Scale for Shine and Opacity
- Graded in shampoo/body wash base

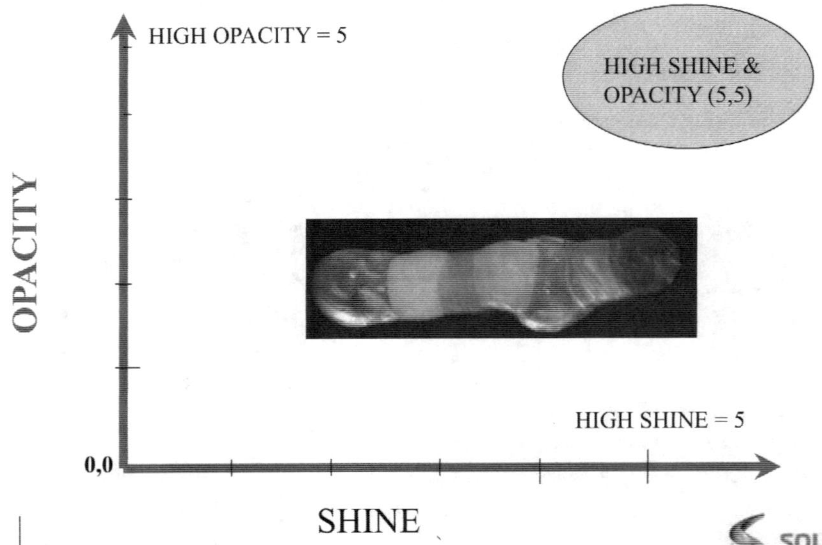

5

Pearl Process Challenges

- Efficient manufacture of high quality cold pearl surfactant blends requires balance of agitation needs for:
  - Heat transfer (heating, cooling/crystallization)
  - Emulsion formation
  - Non-Newtonian fluid flow
  - Formation of desired crystal shape and size
- Compromise on any of these needs can result in reduced quality, process variation, extended cycle times, or an un-scalable process!

6

Emulsion Formation

- Need for axial flow and relatively high shear environment to make/maintain emulsion – Newtonian at this stage

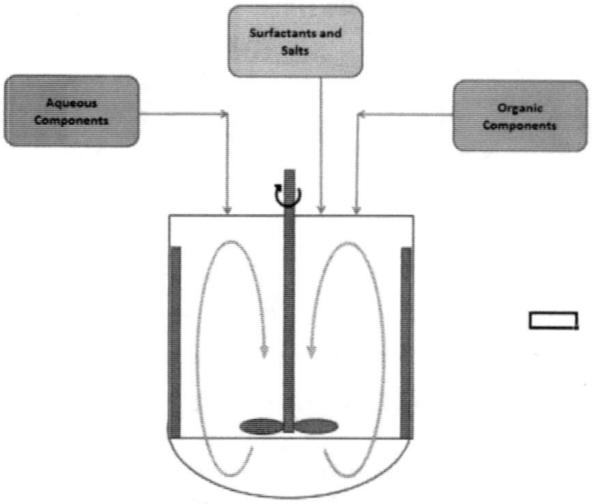

- Composition is a large driver of pearl appearance and emulsion characteristics

7

SOLVAY
asking more from chemistry®

Non-Newtonian Fluid Flow

- Cold pearl surfactant blends exhibit shear-thinning behavior during cooling/crystallization phase – near zero "dead" zone design needed

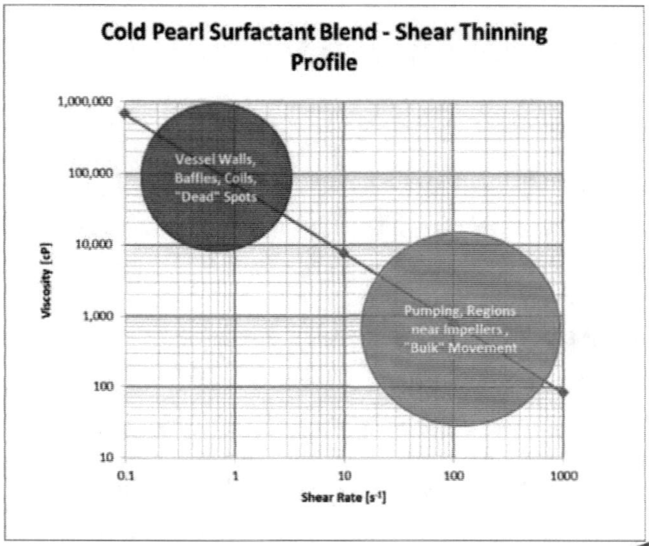

8

SOLVAY
asking more from chemistry®

Crystal Morphology

- Shape and size is affected by applied shear and concentration gradients in crystallizing medium

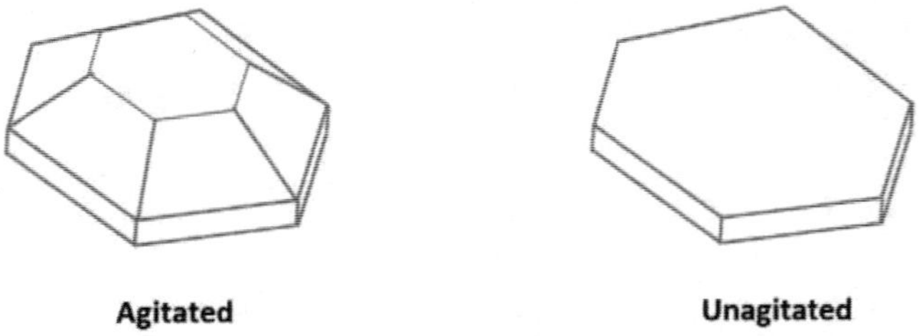

- Increased shear/agitation results in "rounding" of shape, thereby reducing the pearly luster of the material

- The highest quality blends result from extremely low shear environment

9

SOLVAY asking more from chemistry®

Balancing Conflicting Needs

- Simultaneous needs to create emulsion and promote heat transfer, while preserving crystal morphology and product quality
  - Often, compromise is the only solution leading to extended cycle times or lower quality pearl

- Options that do not compromise:
  - Continuous Crystallization – inflexible and somewhat costly (infrastructure)

  - Purpose-built agitation design that addresses **all** needs simultaneously

There are no new problems – look to the food/dairy industry!

10

SOLVAY asking more from chemistry®

Technology Comparison

Traditional Blend Technology
- Single, centrally mounted
- Relatively fast – high shear
- Baffles and jacket cooling – yield loss

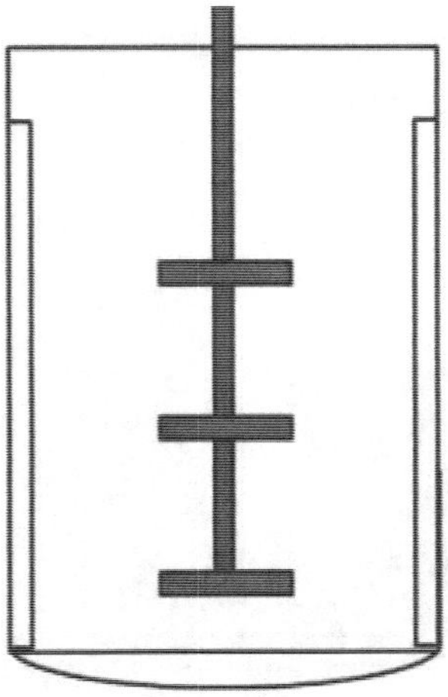

Alternate Blend Technology
- Two counter-rotating, independent
- Wall-wiping, *very* low shear
- Traditional impellers for axial needs

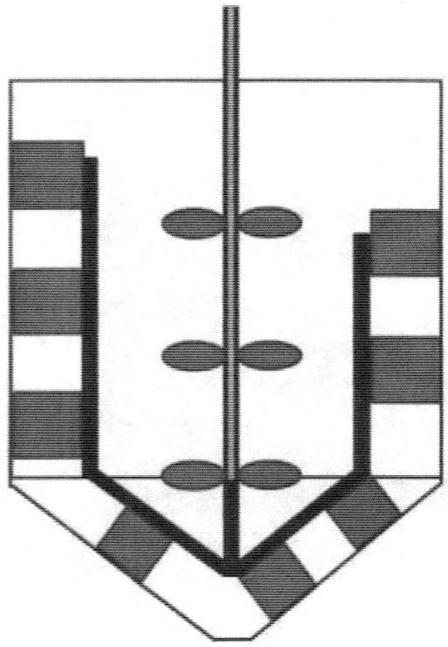

11

SOLVAY
asking more from chemistry®

Pilot Blend Studies

- **A series of pilot studies were commissioned**
- **150 gallon pilot unit rented and installed - dubbed "The Oyster"**
  - **Technical feasibility assessment**
  - **Detailed engineering design study**
- **Define scale-up parameters**
  - **Tip speed?**
  - **Power draw?**
  - **Impact on product quality/cycle time?**

12

SOLVAY
asking more from chemistry®

Pilot Unit Details

- « Zero » Dead Zone Design
- Record of two zone temperatures, fill volume, tip speeds, wattage draws for both impellers and more…
  - Varied process parameters and measured impact on wide array of products
- Empty Video:

13

SOLVAY
asking more from chemistry®

Pilot Study Results

- Validated technology choice
  - Axial elements satisfactory
  - Side-sweeping – zero yield loss bulk movement maintained
- Effect of side-sweeping tip speed identified as critical
- Pearl Video:

14

SOLVAY
asking more from chemistry®

Pilot Study Results

- Tip speed ↗ Opacity ↗
  - •Heat Transfer and shear
- Reduced cycle times, better quality
  - •High Shine/Opacity on grading scale

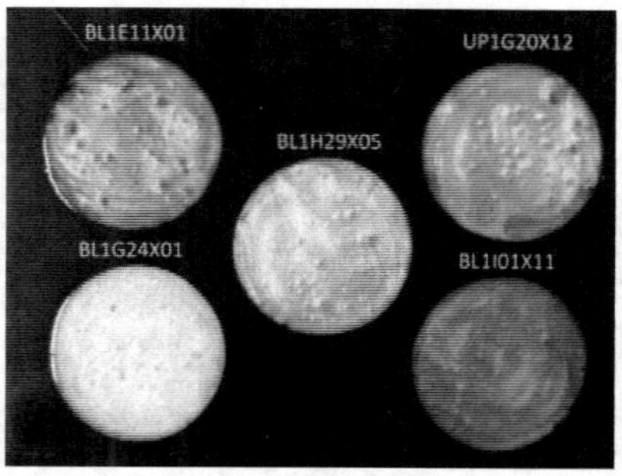

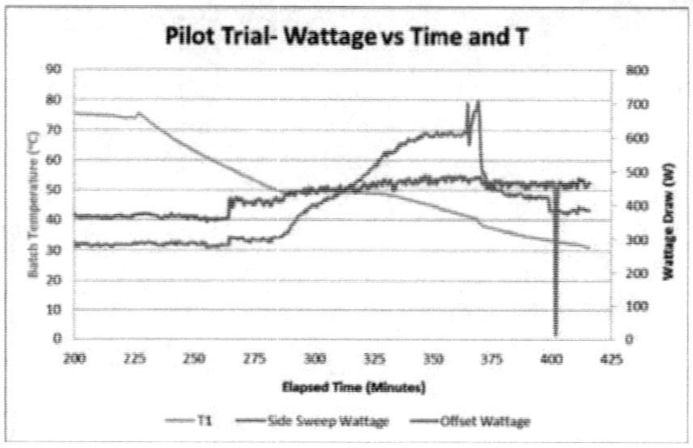

•Engineering Data Collected
- Dynamic power draw for motor sizing
- Tank geometry

•Basis of full scale design

15

SOLVAY
asking more from chemistry®

Industrial Unit Design

- 1:1 scale up of side-sweep tip speed (very scalable)
- High-flow, low-shear axial impellers centrally mounted
- Designed, built and commissioned over 2 years
- Nickname free, so far (Open to suggestions…)

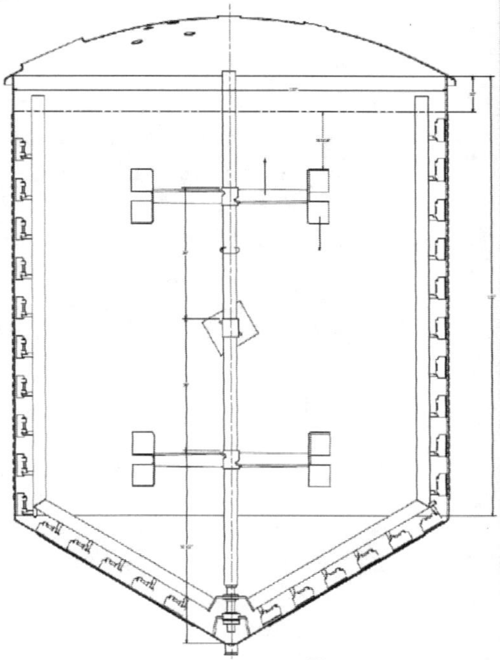

16

SOLVAY
asking more from chemistry®

Industrial Unit Details

- 6000 gallon vessel with cooling jacket
- Optional pump-around loop for axial dispersion and/or external heating/cooling

SOLVAY
asking more from chemistry®

Industrial Commissioning Results

- Appearance of material produced equal to, or slightly above standard in Shine/Opacity with no processing difficulty

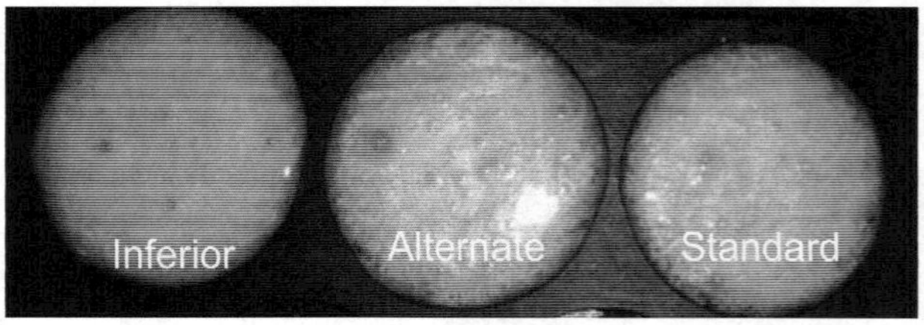

- 50% reduction in cycle time for production of highest quality cold pearl products
- Surprisingly, has benefited production of all performance concentrate blends – reducing cycle times by a factor of 3 in some cases!
  - Due to "zero" dead zone design, homogeneity is achieved very quickly in shear-thinning systems

SOLVAY
asking more from chemistry®

Conclusions

- Cold Pearl Surfactant Blends present varied and unique challenges in their manufacture
- The alternate agitation design described here today has:
  - Simultaneously solved the need to maintain the desired product quality, while ensuring sustainable process capability
  - Offered **a** unique solution to a complex problem
- Benefits to pearls and other products have been many
  - Reduced cycle times
  - Increased quality
- Allows for a highly scalable R&D path – just the start of intelligent blend design!

19

SOLVAY
asking more from chemistry®

www.solvay.com

SOLVAY
asking more from chemistry®

VIRGINIA COMMONWEALTH UNIVERSITY

School of Engineering

AIChE Process Development Symposium

"Pilot Plants" session , June 13, 2013:

Successful Scale-up and Piloting -

Overcoming Challenges Other

than Obvious Technical Issues

Successful Scale-up and Piloting -
Overcoming Challenges Other than Obvious Technical Issues

Lessons learned from 30 years of process development and production support

or

It's good to learn from your mistakes, but it's better to learn from others' mistakes

VCU School of Engineering
VIRGINIA COMMONWEALTH UNIVERSITY

Successful Scale-up and Piloting -
Overcoming Challenges Other than Obvious Technical Issues

Rudolph (Rudy) Krack career:

- VCU Chemical and Life Science Engineering 2008-present
- Boehringer Ingelheim Chemicals (BICI) * 2000-2007
- Boehringer Ingelheim Pharmaceuticals (BIPI) 1997-2000
- Novartis (formerly Ciba-Geigy) 1996-1997
- Hoffmann LaRoche * 1991-1996
- Nepera Chemicals * 1977-1991
- J. T. Baker Chemicals * 1975-1976

* Chemical plant sites with a development group and a pilot plant. These 4 sites produced active pharmaceutical ingredients (APIs).

VCU School of Engineering
VIRGINIA COMMONWEALTH UNIVERSITY

Successful Scale-up and Piloting -
Overcoming Challenges Other than Obvious Technical Issues

The Pilot Plant's role in Process Development:

- Transition from lab bench to production.
- Bridge from R&D to manufacturing.
- One step on the path to commercialization.

VCU School of Engineering
VIRGINIA COMMONWEALTH UNIVERSITY

Successful Scale-up and Piloting - Overcoming Challenges Other than Obvious Technical Issues

Stage / Gate Process for:
Process Development / Scale-up / Transfer to Production

From VCU course ENGZ 402 Senior Design Studio Lab

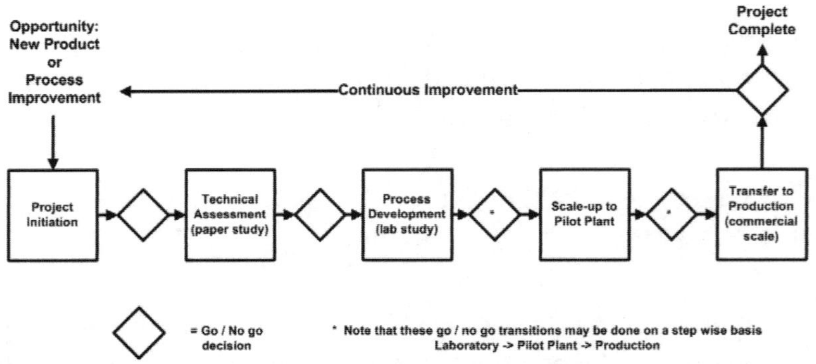

VCU School of Engineering
VIRGINIA COMMONWEALTH UNIVERSITY

Successful Scale-up and Piloting - Overcoming Challenges Other than Obvious Technical Issues

Pilot Plants, typical features from RK experience:

- Smaller version of production facility; similar equipment, materials of construction, and process control systems
- Usually reports to Research and Development
- Staffed by mix of chemists / engineers with degree(s) and experienced operators with or without degree(s)
- Often used for routine, small scale production
- Scale-up factors can be 100 – 5000 from the laboratory bench, 10 – 100 to production facility

VCU School of Engineering
VIRGINIA COMMONWEALTH UNIVERSITY

Successful Scale-up and Piloting -
Overcoming Challenges Other than Obvious Technical Issues

R. Krack Observations:

- Not all piloting is done in pilot plants.
  Overlapping equipment scales:
    Lab (10 mL-1000 gal), Production (10 L-10,000 gal)
    Multiple scale-ups are possible in each area.
- Pilot plants don't always do piloting.
  Often used for routine, small scale production.
- Information gained during piloting is often more important than the product made in the pilot plant.

VCU School of Engineering
VIRGINIA COMMONWEALTH UNIVERSITY

Successful Scale-up and Piloting -
Overcoming Challenges Other than Obvious Technical Issues

R. Krack Observations:

- Consolidation led to centralized development and pilot plants.
  Fewer production facilities have these capabilities on-site.
- Increased specialization of process development activities.
  Have site specific roles in overall development process.
- Greater use of outsourced / off-site providers for all phases of process commercialization: laboratory, pilot and production.
- Piloting requires even more attention to interaction among various departments, sites, companies and even countries.

VCU School of Engineering
VIRGINIA COMMONWEALTH UNIVERSITY

Successful Scale-up and Piloting -
Overcoming Challenges Other than Obvious Technical Issues

Technology transfer is not just sending documents to the receiver.

It is the transfer of process knowledge.

(paraphrase of one of the opening statements at an ISPE New England chapter session on Technology Transfer at BIPI in the late 1990s)

VCU School of Engineering
VIRGINIA COMMONWEALTH UNIVERSITY

Successful Scale-up and Piloting -
Overcoming Challenges Other than Obvious Technical Issues

Nepera Chemicals (NY):

- Site active from 1940s until 2003
- Initially privately owned into 1950s
- Owned by Warner-Lambert 1950s -1976
- Owned by Schering AG 1976 to 1987
- Owned by Cambrex (nee CasChem) 1987 to 2003
- Multi-product, batch API facility & pilot plant built in 1960s:
  - Open floor plan, no separation of "bays", except for high pressure cubicles, i. e. for hydrogenation
  - Multiple utility supplies to each vessel: steam, cooling water, chilled brine; temperature manually controlled

VCU School of Engineering
VIRGINIA COMMONWEALTH UNIVERSITY

Successful Scale-up and Piloting -
Overcoming Challenges Other than Obvious Technical Issues

Hoffmann LaRoche (NJ):

- Site active from 1920s until present
- Multi-product, batch API facilities built from 1940s to early 1960s
- Pilot plant built in late 1950s-1960
- Open floor plan, equipment grouped to "bays"
- Multiple utilities to each vessel, local control
- Selected retrofits to upgraded control systems

VCU School of Engineering
VIRGINIA COMMONWEALTH UNIVERSITY

Successful Scale-up and Piloting -
Overcoming Challenges Other than Obvious Technical Issues

Boehringer Ingelheim Chemicals, Incorporated (BICI):

- Site active from 1979 to the present for API production; batch, multi-product.
- Originally a 50-50 joint venture of A. H. Robins and BI. BI became 100 % in 1991.
- Multi-product, batch, active pharmaceutical ingredient production facility
- Original building had open floor plan, equipment grouped into "bays"
  - Multiple utilities to each vessel: local temperature control
  - Selected retrofits to upgraded control and single jacket fluid
  - Smallest reactors used as "pilot plant"
- New "Specialty" building with pilot plant and launch bay (1998).
  - Single fluid jacket utility system, distributed process control system
  - Kilo lab: glass reactors in fume hood for potent compound handling (1999)
  - Expansion to production bays in 2000s
- New manufacturing building with upgraded (enclosed) material handling (2008)

VCU School of Engineering
VIRGINIA COMMONWEALTH UNIVERSITY

Successful Scale-up and Piloting - Overcoming Challenges Other than Obvious Technical Issues

Management decisions can overrule the planned scale-up execution (1).
At Nepera, a normal scale-up and piloting process was sent on a tangent:

- Assigned to project when chemist almost finished with bench work.
- With chemist, manufactured a pilot batch.
- Developed capital cost estimate to retrofit existing facility.
- ***Capital estimate exceeded capital plan budget.***
- ***Reworked estimate to allocate costs: new product, environmental upgrade, building upgrade, and previous product made in that building.***
- ***Charged by management to produce new product in existing building.***
- ***With two chemical operators who previously worked in building, spent 2 weeks working 24 / 7 with chemist to manufacture one batch.***
- Management approved 3 of 4 capital projects, not for old product.
- Executed project(s), started up process, resolved several scale-up issues.

VCU School of Engineering
VIRGINIA COMMONWEALTH UNIVERSITY

Successful Scale-up and Piloting - Overcoming Challenges Other than Obvious Technical Issues

Management decisions can overrule the planned scale-up execution (2).
At Boehringer Ingelheim, lost new launch facility to routine production:

- Negotiations for BICI as second supplier.
- Production at client's facility disrupted.
- Transfer to new launch facility at BICI expedited.
- First batch of new product was the first batch in this facility.
- ***Lost the intended use of the new launch facility for new product scale-up.***
- ***Launch facility became a dedicated production.***

VCU School of Engineering
VIRGINIA COMMONWEALTH UNIVERSITY

Successful Scale-up and Piloting - Overcoming Challenges Other than Obvious Technical Issues

Even the most trivial omitted details may be important (1).
At Nepera, communication failures during scale-up from pilot plant:

- Finished bench work, then chemist, technician and engineer made a pilot batch.
- Process transferred to production; used existing equipment, no capital spent.
- Initial batch had only 50% of expected yield.
- ***Temperature control: lab batches were 5 ˚C +/- 0.1 ˚C.***
  Plant batch was 5 ˚C +/- 5 ˚C due to manual control of cooling.
- ***Caustic addition temperature: lab made up 50% NaOH , cooled before addition.***
  Plant supply via a steam traced pipeline from a heated outdoor tank.
- Production procedure had stated: ***"Control at 5 ˚C"*** and ***"add 50% NaOH"***.
- Executed small capital project: jacket recycle pump and temperature controller.
- Added instructions to cool 50% NaOH to < 10 ˚C before addition.

VCU School of Engineering
VIRGINIA COMMONWEALTH UNIVERSITY

Successful Scale-up and Piloting - Overcoming Challenges Other than Obvious Technical Issues

Even the most trivial omitted details may be important (2).
At BICI, communication issues during transfer between sites:

- Initial small scale campaign at Process Development Laboratory at BIPI CT.
- ***Last campaign in CT was completed about one 1 year before assignment to transfer this product to BICI VA.***
- Months of document transfer from BIPI CT to BICI VA, laboratory development work at BICI VA, safety reviews and other planning.
- ***First batch run in the BICI pilot plant also the first batch run in this equipment.***
- First reaction is a chlorination with HCl gas under very low positive pressure.
- ***The HCl cylinder used was equipped with a dip tube and delivered HCl liquid.***
- ***As a result, material was over-chlorinated.***
- Added specification for HCl cylinder without dip tube.

VCU School of Engineering
VIRGINIA COMMONWEALTH UNIVERSITY

Successful Scale-up and Piloting - Overcoming Challenges Other than Obvious Technical Issues

Development activities at multiple sites increases communication risks. (1)
At Nepera, communication failures between multiple sites:

- Laboratory bench work done outside the corporation, possibly a university.
- Pilot plant batch done at NJ site of sister company, observed by NY engineer.
- ***Safety review for plant batch at NY site identified a problem with use of dry catalyst (precious metal on carbon) for hydrogenation.***
- ***Decision made to pre-wet catalyst with the key raw material at NJ site.***
- Hydrogenation required higher pressure than used for normal production.
- Despite the higher hydrogen pressure, there was no hydrogen uptake.
- ***Investigation revealed that the raw material deactivated the catalyst during the time between wetting in NJ and use in the NY plant hydrogenation reactor.***

VCU School of Engineering
VIRGINIA COMMONWEALTH UNIVERSITY

Successful Scale-up and Piloting - Overcoming Challenges Other than Obvious Technical Issues

Development activities at multiple sites increases communication risks. (2)
At Nepera, communication failures between multiple sites:

- Laboratory bench work done at client sites ***(VA and RI)*** and at Nepera ***(NY).***
- Pilot plant campaign at a different client site ***(TX).***
- Production at a sister company ***(PA),*** management and QC from client ***(VA).***
- Chemists and engineering support from sister companies ***(NJ, NY, PA).***
- Process Equipment testing at vendor sites ***(IL an NC).***

VCU School of Engineering
VIRGINIA COMMONWEALTH UNIVERSITY

Successful Scale-up and Piloting - Overcoming Challenges Other than Obvious Technical Issues

Development activities at multiple sites increases communication risks. (3)
At BICI, communications issues within a multi-national company:

- Laboratory development, piloting and launch in Germany.
- Transfer to US production at BICI VA.
- Observation of production in Germany by BICI visitors.
- Limited pilot plant campaign at BICI VA.
- Personnel from Germany observe and support initial BICI batches.
- ***Numerous transfer issues including: differences in language and culture,***
- ***primarily chemists versus mix of chemists and engineers,***
- ***company headquarters versus plant site.***
- ***Small but important detail; European number and date conventions:***
- ***Day / month / year rather than month / day / year,***
- ***Comma and period usage. In Europe, 3.000 = 3 thousand.***

VCU School of Engineering
VIRGINIA COMMONWEALTH UNIVERSITY

Successful Scale-up and Piloting - Overcoming Challenges Other than Obvious Technical Issues

Inadequate in-process analytical methods results in trouble later.
At Nepera, passed in-process analysis method, but failed later QC analysis.

- Laboratory and initial pilot work done at client's development site in NY.
- Next phase of scale-up done at NY facility, using largest batch reactors on site.
- Ultimate production at client facility in AL.
- Batch reaction completion test done at client's NY site.
- Batch filtration wash test done at contract manufacturer's NY site.
- ***Subsequent tests at client's AL site determined that the wash was inadequate.***
- ***Filtration wash in-process test was therefore also inadequate.***
- ***Material was re-washed at client's AL site.***
- ***Financial adjustment to contract payment to cover cost of client re-processing.***

VCU School of Engineering
VIRGINIA COMMONWEALTH UNIVERSITY

Successful Scale-up and Piloting - Overcoming Challenges Other than Obvious Technical Issues

Acknowledgements:

Frank Gupton Ph. D.: Chair, VCU Chemical and Life Science Engineering; was Executive Director, Process Development, Boehringer Ingelheim Chemicals 1993-2005

Laurence J. (Larry) Nummy Ph.D.: Senior Research Fellow, Chemical Development, Boehringer Ingelheim Pharmaceuticals; was Senior Research Chemist, Nepera Chemicals 1982-1990

Ken Stanley: Specialist, Process Chemistry; Boehringer Ingelheim Chemicals

Ross Topliff: Principal Tops Engineering; Pilot Plants session chair 2013; was Process Engineer, Nepera Chemicals 1980-1987

VCU School of Engineering

VIRGINIA COMMONWEALTH UNIVERSITY

The TRI Integrated Biorefinery Pilot Plant

Ravi Chandran[1]
Dave Newport[1]
Kym Arcuri[2]
Sean Whitney[1]
Daniel Leo[1]
Shawn Freitas[1]

[1] TRI , [2] EFT

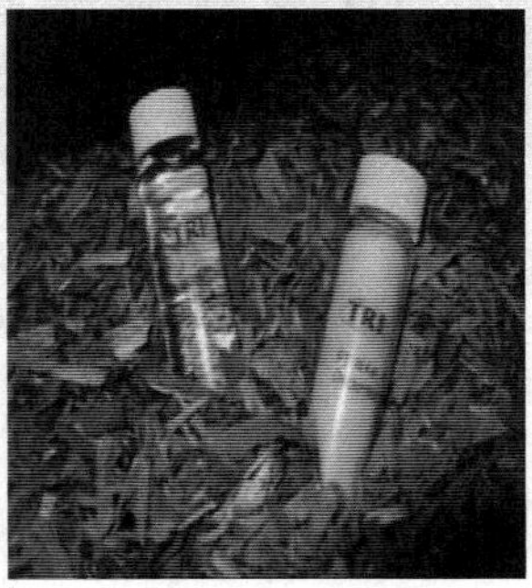

2013 AIChE Process Dev. Symposium, June 13, 2013

TRI

TRI's Core Technology

TRI

- Deep fluidized bed
- Proprietary indirectly-heated steam reforming process at medium temperature and low pressure
- 3 inputs and 3 outputs
- Biomass undergoes drying, devolatilization, water gas, hydrogenation and steam reforming

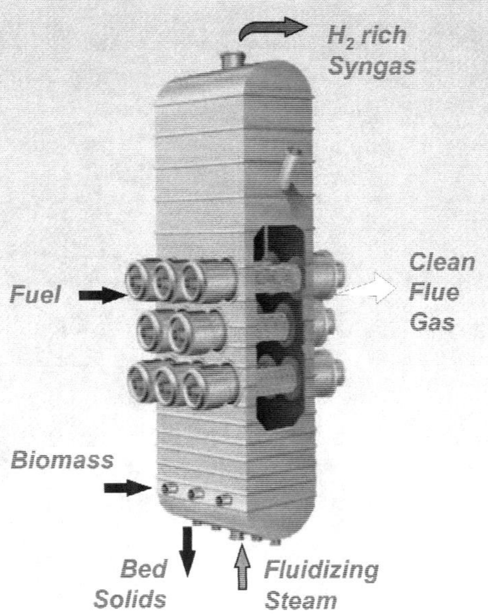

www.tri-inc.net

2

Technology Development Chronology

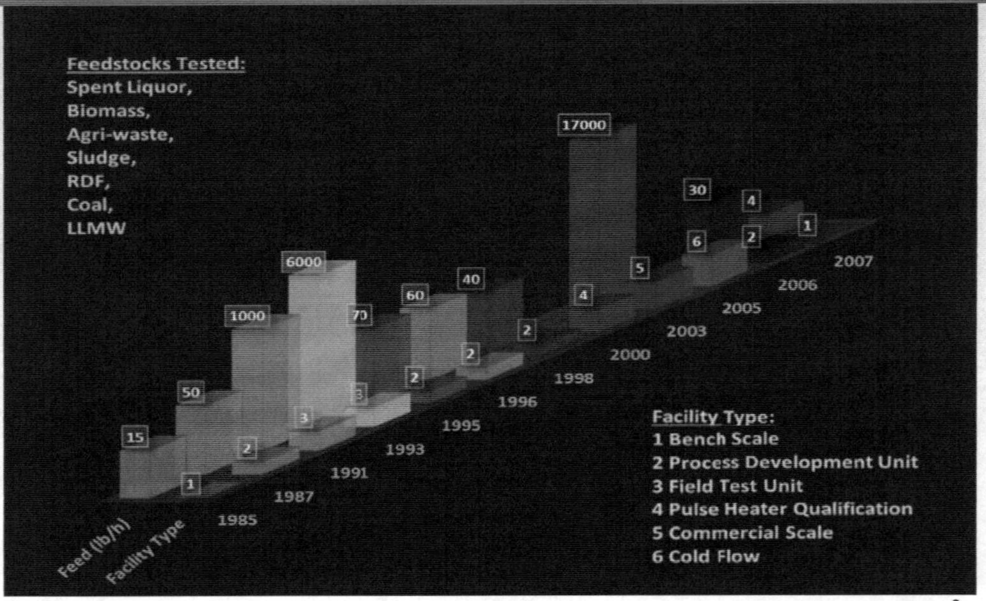

www.tri-inc.net

3

Scale-up Experience

Black Liquor

60
pph

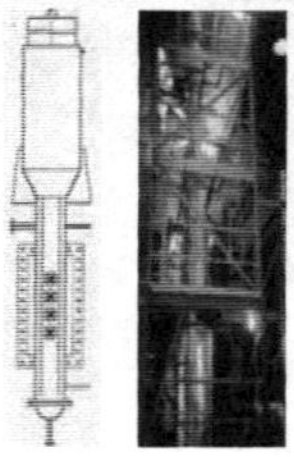

17,000
pph

www.tri-inc.net

4

Pilot Plant Objective

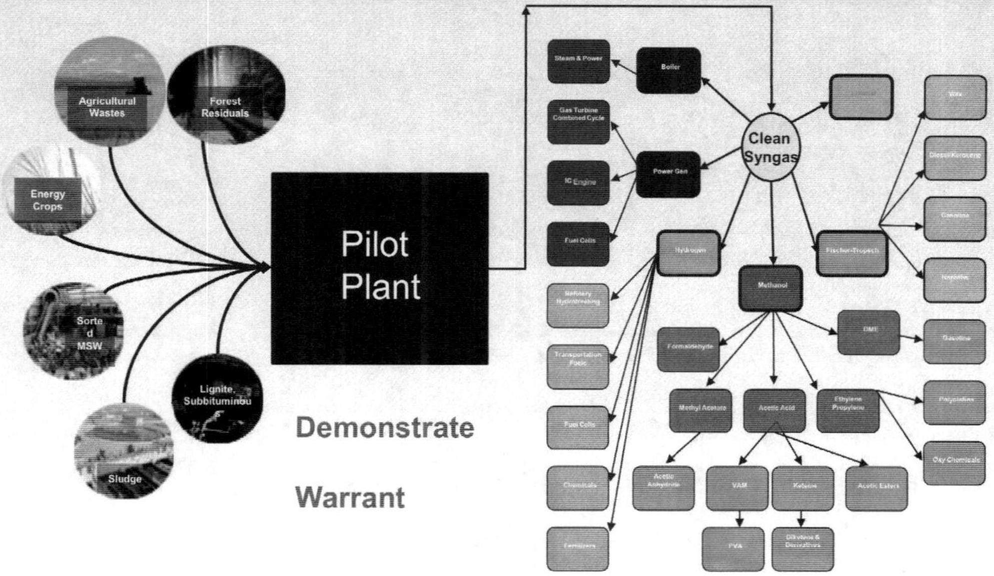

www.tri-inc.net

5

Pilot Plant Design

- Feedstock flexibility
  - Physics (size, shape, density)
  - Chemistry (organics, HHV, inorganics, contaminants)
- Many reactions – *drying, devolatilization, water gas, hydrogenation, gasification, partial oxidation*
  - Quantitative and qualitative factors
  - Unit specific and feedstock specific parameters
- Scaling
  - Physics of fluidization or bed fluidization dynamics
  - Chemistry of feedstock
- Syngas cleanup and conditioning
- Syngas customization

www.tri-inc.net

6

TRI Process Configuration

- Unique 2-stage gasification
- High quality, medium calorific value syngas
- Customized syngas composition specific to downstream processing
- Feedstock flexibility
- High carbon conversion
- Thermal integration
- Inherently stable and safe
- Scalable

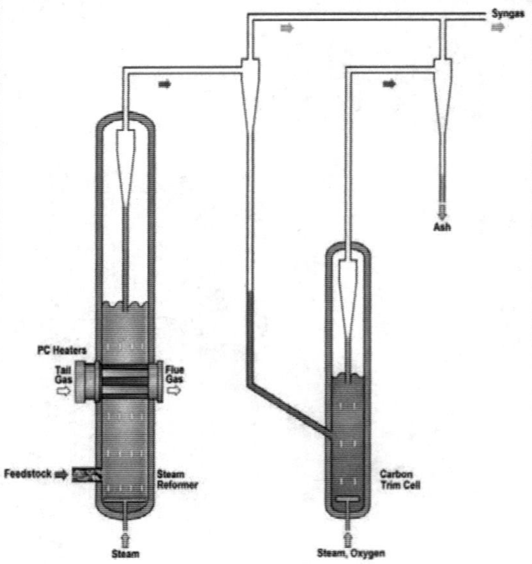

www.tri-inc.net

7

Own or Outsource?

- Build, own and manage
- Contract R&D
  - 6 candidates
  - 2 finalists
  - Selected a facility in Durham, NC

www.tri-inc.net

8

Pilot Plant Initial Concept

www.tri-inc.net

9

TRI Integrated Biorefinery Pilot Plant

- Capacity: 4 dtpd biomass, 1 MW thermal
- Scope
  - *Initial*
    - Biomass feed
    - Biomass gasification
    - Primary syngas clean-up (PGCU)
  - *Modified to include*
    - Syngas compression
    - Secondary syngas clean-up (SGCU)
    - F-T liquids production (FTL)
  - Every single unit operation in commercial system represented

www.tri-inc.net

10

Pilot Plant Implementation

- Material and Energy Balances
- Process flow diagram
- Equipment list and specifications
- Design review
- Data sheets
- Piping and instrumentation diagrams
- Motor list
- Equipment layout
- Environmental impact
- Procurement
- Equipment selection
- Installation
- Hazop analysis
- Functional control specifications
- Standard operating procedures
- Analytics
- Commissioning
- Shakedown
- Operation

www.tri-inc.net 11

Pilot Plant Block Flow Diagram

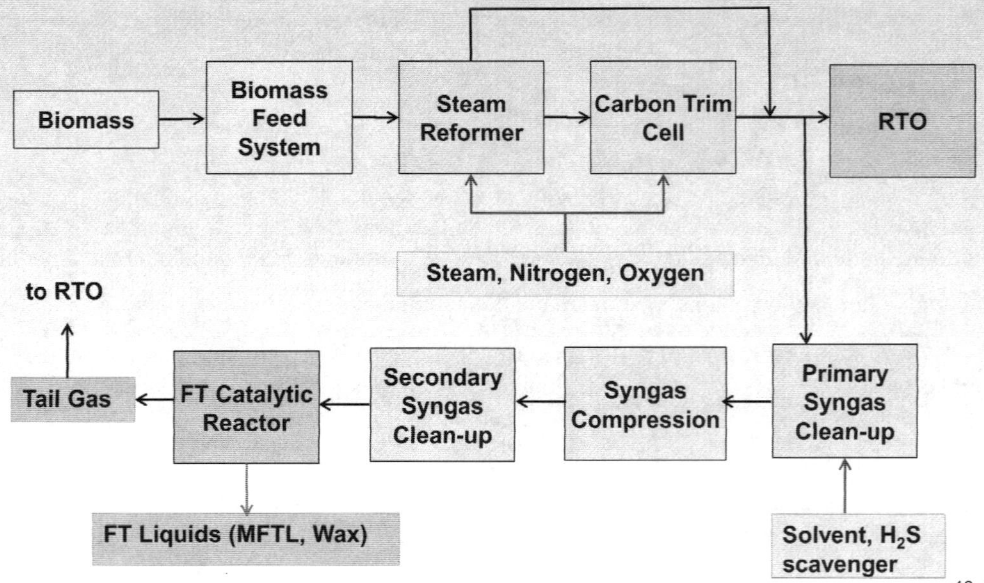

www.tri-inc.net 12

TRI Integrated Biorefinery Pilot Plant

Test Objectives

- Verify Performance and Operability
  - Syngas composition and yield
  - Maintainability at the desired H_2 to CO molar ratio
  - Tar yield and composition
  - Carbon conversion
  - Endothermic heat of reaction
  - Syngas contaminant levels after PGCU
  - Syngas cleanliness upstream of F-T
  - CO conversion in the F-T, FTL yield
  - Thermal stability of F-T during transient operations
  - Catalyst activity, selectivity and deactivation rate
  - Effluent generation rate and its characteristics
  - Mass and Energy balance
- Support Commercial Unit design

Yin and Yang / Agony and Ecstasy

- Operational
- Measurement
- Gas cleanliness
- Schedule
- Budget

- **100% renewable fuel**
- **Met or exceeded all the targets**

www.tri-inc.net

15

Feedstock

Description	Value
Feedstock	Woody biomass
Origin	NP, FRB wood basket area
Biomass Moisture Content	Avg 8 wt%; range 6 – 22 wt%
<u>Ultimate Analysis</u> C H N S O Ash	<u>Wt%</u> 52.18 5.53 1.09 0.01 40.48 0.71
HHV	19.85 MJ/kg dry (8,532 Btu/lb dry)
Feed size	12 x 12 x (12-37) mm
Feed bulk density	Avg 200 kg/m^3 (12.5 lb/ft^3)

www.tri-inc.net

16

Technical Performance – Pilot Plant

Description	Value
Biomass Feed Rate	Avg 4 tpd dry; range 2 to 5
Carbon conversion (gasifier)	93 - 98%
H_2 to CO ratio	1.75 - 2 (for stage 1 F-T); 1.45 -1.7 (for stage 2 F-T)
Syngas composition	**Vol% (dry basis)**
H_2	28.8
N_2	30.4 (due to scaling; 1.5-2 in commercial)
CH_4	7.0
CO	15.2
CO_2	17.2
C_xH_y	1.4
HHV	If N_2 is1.8%: 12.5 MJ/m^3dry(335 Btu/dscf)
Primary Tars Condensable tars	1.8 – 5.4 g/Nm^3 (325 - 970 ppmv) 0.06 – 0.90 g/Nm^3 (9 -150 ppmv)
Contaminants in syngas from PGCU	< 10 ppmv

www.tri-inc.net

Technical Performance – Pilot Plant (continued)

Description	Value
Contaminants in syngas from SGCU	Below detection limit or < 10 ppbv
F-T CO conversion (single pass)	60+% - 1st stage mode with N_2 diluted syngas 65% - 2nd stage mode
F-T Liquids	100% renewable fuels - pure white wax (C21+) - water clear medium fraction HC liquids
F-T Catalyst productivity, activity, methane selectivity	Met or exceeded design goals
Byproduct	Clear BTEX liquid
Wastewater effluent	Non-hazardous COD 280 – 500 mg/L TSS 1 – 58 mg/L

18

www.tri-inc.net

Technology Key Accomplishments

TRI

Pilot Demonstration Successfully completed

- Integrated Biorefinery trial met or exceeded performance criteria
- Generated high quality F-T products for off-take agreements
- Non-detect level of contaminants after gas clean-up
- No observable F-T catalyst deactivation

Extensive Run Time

- 7,500 hrs on gasification
- 4,000 hrs on F-T GTL
  - 2,800 hrs on biomass syngas
  - 1,200 hrs on trailer syngas

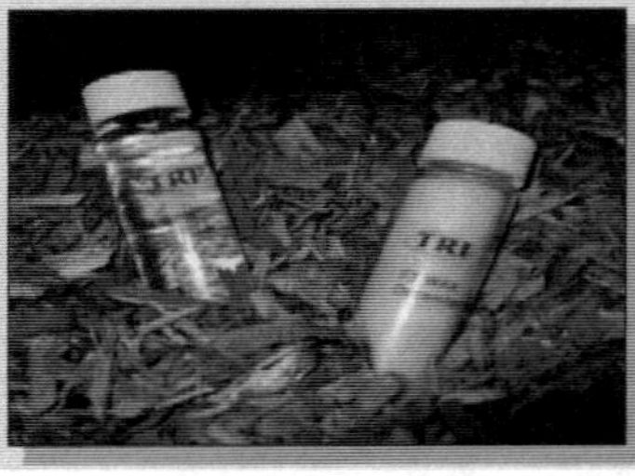

www.tri-inc.net

19

TRI Pilot Plant Summary

TRI

Stable Consistent Gasification

- Reliable feed system
- Efficient two stage gasification with low tars generation & consistent H_2:CO ratio

Effective Primary and Secondary Gas Clean-up (KASyn)

- Non-detect levels of all contaminants – at or below ppb threshold for analytical methods (Enhanced analytical methods implemented)
- Recovered Benzene, Toluene and Xylene (BTX) as clean byproduct

Efficient Conversion of Syngas to F-T Fuels and Waxes

- High CO conversion rates with good catalyst productivity
- High quality F-T products generated for off-take agreements
- Very low catalyst deactivation rates observed – good catalyst life
- Zero contaminants detected on post-trial catalyst elemental analysis

www.tri-inc.net

20

Conclusions

www.tri-inc.net 21

3D Printed Pilot Plant Next?

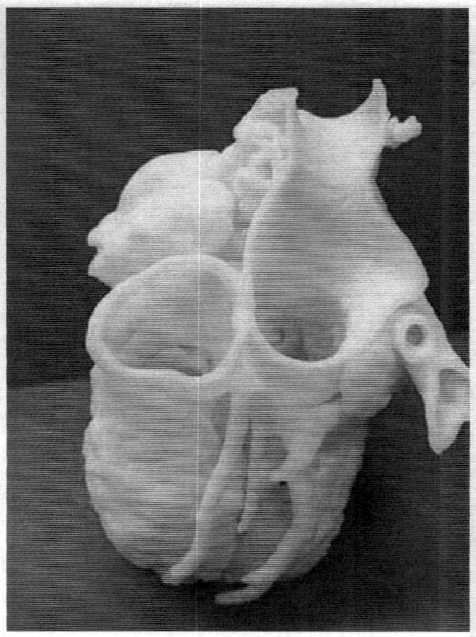

www.tri-inc.net 22

Filtration, Cake Washing & Drying: Lab Testing to Pilot Testing to Project Installation and Performance Guarantees

Barry A. Perlmutter,
President & Managing Director
BHS-Filtration Inc.
9123-115 Monroe Road
Charlotte, North Carolina 28270

BHS FILTRATION

BHS-Inc.

Thin-Cake Solid-Liquid Separation, Cake Washing & Drying Technologies

- Pressure & Vacuum Filtration
- Batch & Continuous Operation
- High Solids to Clarification
- Cake Washing & Drying
- Automatic Discharge of Wet Cake, Dry Cake or Concentrated Slurry

BHS FILTRATION

BHS-Inc.

Thin-Cake Solid-Liquid Separation, Cake Washing & Drying Technologies

Candle Filter:	Clarification
Pressure Plate Filter:	Clarification
Vacuum Belt Filter:	High Solids, Continuous
Rotary Pressure Filter:	High Solids, Continuous
Autopress Filter:	Full Containment, Pharma

Main Applications: chemical, energy, environmental, mining & pharmaceutical

BHS FILTRATION

Presentation Overview

- **Generally, after the initial mixing - reactions, there is bulk separation to remove the larger product solids or unwanted solids.**
- **There is a trend to remove the residual particle fines from the slurries.**
- **These fines are very small, typically in the 1 – 5 micron range and smaller and are very low in concentration down to ppm levels.**

BHS FILTRATION

Presentation Overview

- Solving a filtration problem:
  - Lab testing, Pilot testing & Scale-Up,
  - Installation & Performance Guarantees
- The overall scheme can be used by process engineers to develop optimum filtration solutions for pressure or vacuum, batch or continuous and high-solids to clarification applications.

BHS FILTRATION

Overview of Current Process

- The current process filters a specialty chemical in a filter press with paper filter media at 170 °C.
- Diatomaceous earth and/or activated carbon were used as a body feed and precoat.
- The process had typical problems associated with a filter press including bypass of slurry, manual operation, filter media plugging, non-reproducible quality plus the safety concerns of hot oil.

BHS FILTRATION

Laboratory Tests

BHS-Pocket Filter

20 cm^2 filter area

400 ml content

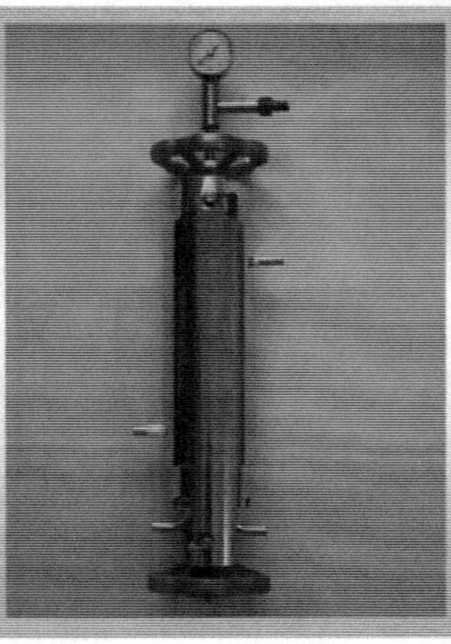

Laboratory Tests: Pressure

Cracking/Shrinking & Optimum Cakes

BHS FILTRATION

SELECTION OF FILTRATION TECHNOLOGY

CAKE STRUCTURE /PROCESS PARAMETERS FROM LAB TESTING

- **Cake Thickness and Filtration:**
- **Filter Media**
- **Cake Washing**
- **Precoat / Body Feed**
- **Cake Drying**
- **Cake Discharge**
- **CIP/SIP**

BHS FILTRATION

SELECTION OF FILTRATION TECHNOLOGY

Run 4: Filter Media Polypropylene (8 um) with DE precoat

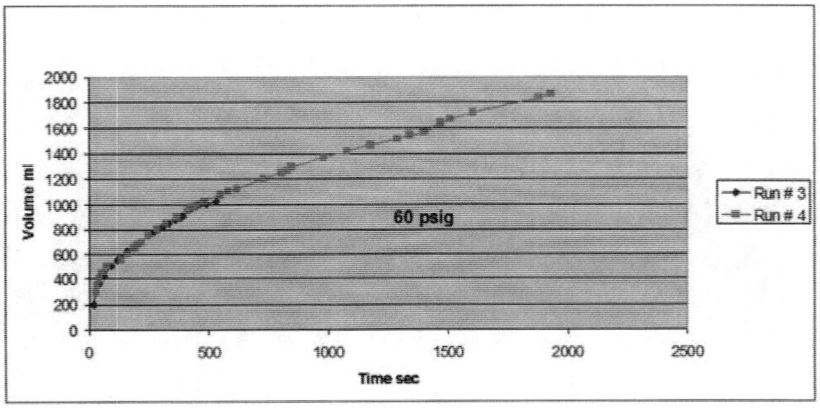

SCALE-UP & SELECTION OF FILTRATION TECHNOLOGY

Filling	5 Minutes
Pre-coating	10 Minutes
Filtration	110 Minutes
Draining	5 Minutes
Drying	15 Minutes
Discharge	10 Minutes
Total Processing Time	155 Minutes (Required Rate: 5677 liters/3 hours)

- **Test Filter Performance:**
  $(1.865\ l / 20\ cm^2)/\sqrt{32}\ min \times 10000\ cm^2 / m^2 = 164.3\ l/m^2 \sqrt{min}$
- **Specific Rate Test Filter: $164.3\ l/m^2 \sqrt{min} \times \sqrt{110}\ min. = 1723\ l/m^2$**
- **Filter Area = Required Rate / Specific Test: $5677.5\ l / 1723\ l/m^2 = 3.3\ m^2$**
- **BHS Candle Filter: CF 18-007 provides a filtration area of $3.56\ m^2$**

Batch/Continuous Candle Filters for Low Solids Concentrations

- Applications include water & amine scrubbers, chlor-alkali, chemical, biodiesel, gasification / grey water, activated carbon, catalysts, heavy metals, PTA fines and pharmaceutical
- Replaces filter presses, manual plate filters, bag, cartridge, metal filters and centrifuges
- Filtration to 0.5 um, cake washing and drying
- Precoat of the filter media is possible, if required
- Complete containment and Automatic discharge

BHS FILTRATION

Candle Filter Technology

BHS FILTRATION

Candles and Registers

BHS FILTRATION

Detail of the BHS Candle

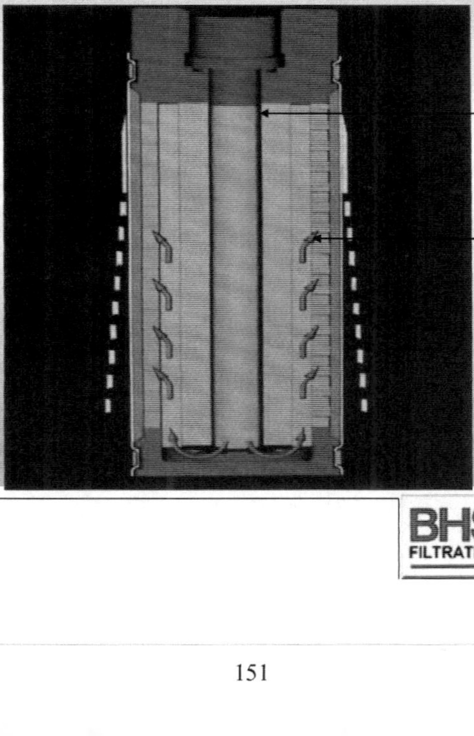

Filtrate Pipe

Gas Flow Shown in Purple

BHS FILTRATION

BHS Candle – Filter Media Sock

BHS Rental Candle Filter CF012-3 (1.0 m²) of filter area

Test device: Candle Filter CF 012-3 with 1 m² of area
Filter cloths: Teflon material: 5 um pore size
PEEK: 22 um pore size
Temperature: Heated slurries, 150 degrees C (max)
Drying: Compressed air at ambient temp
Blowback: Compressed air at ambient temp

BHS Rental Candle Filter CF012-3 (1.0 m^2) of filter area

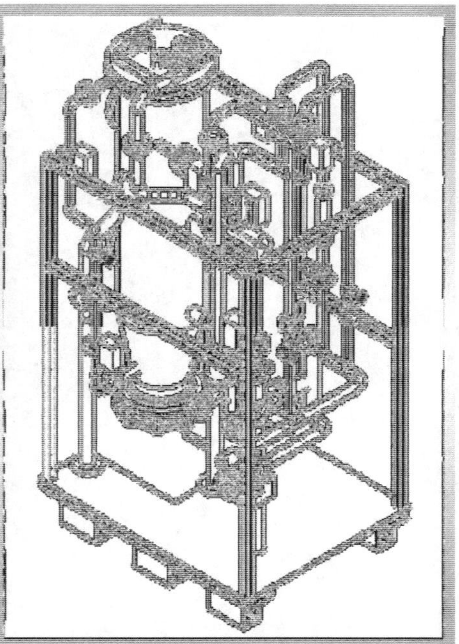

BHS FILTRATION

PILOT TEST SCALE-UP

Filling	5 Minutes
Pre-coating	10 Minutes
Filtration	87 Minutes
Draining	10 Minutes
Drying	5 Minutes
Discharge	10 Minutes

Total Processing Time: 137 Minutes

Required Rate: 5677 liters/3 hours

- **BHS Candle Filter:**
- **CF 18-019 provides a filtration area of 10 m^2**

BHS CANDLE FILTER, CF 18-019 WITH PLC CONTROLS

BHS CANDLE FILTER PERFORMANCE GUARANTEE

A) BHS conducted tests using the BHS CF 12-003. The specific test report is the basis for the scale-up and process guarantee.

B) Based upon the above referenced, mutually agreed test results, BHS will supply one candle filter system consisting of a CF 18-019 with PTFE socks.

C) MECHANICAL GUARANTEE:

D) QUALITY GUARANTEE:

E) VOLUME PERFORMANCE GUARANTEE:

F) FILTER SOCK PERFORMANCE GUARANTEE:

G) FINAL PRODUCT QUALITY SPECIFICATIONS:

Product	Color Spec	Sediment mg/L	Sediment Ash, mg/L	Sebacic Sediment, mg/L	Filter Time min/L	Ash wt%
600						
CW						
F						
FABP						
J						
7H						
22H						
32H						
68H						

Summary

- There are many choices for filtration in a process.
- Technical evaluation, laboratory and pilot testing are critical for successful decisions and projects.
- Engineers must evaluate all outcomes to make an informed and successful decision
- The take-away is that close collaboration between the operating company and the vendor will allow for creative problem-solving and process filtration solutions to achieve the desired quality and production requirements

BHS FILTRATION

THANK YOU!

Barry A. Perlmutter
President & Managing Director
14300 South Lakes Drive
Charlotte, North Carolina 28273
Tel: 704.845.1190; Fax: 704.845.1902
E-mail: info@bhs-filtration.com

BHS FILTRATION

United States Environmental Protection Agency

Design of Sustainable Energy Supply Chains using the P-Graph methodology involving various metrics

Leisha Vance[1], Istvan Heckl[2], Botond Bertok[2], **Heriberto Cabezas**[1], and Ferenc Friedler[2]

[1]U.S. Environmental Protection Agency, Office of Research and Development, Cincinnati, Ohio, USA
[2]University of Pannonia, Department Computer Science and Systems Technology, Veszprém, Hungary

Office of Research and Development
National Risk Management Research Laboratory, Sustainable Technologies Division

June 20, 2013

United States Environmental Protection Agency

Outline

- Sustainability
- P-Graph Framework
- Sustainability Metrics
- Supply Chain Structures
- Scale-Up
- Summary

1

United States Environmental Protection Agency

SUSTAINABILITY

2

United States Environmental Protection Agency

Sustainability Metrics

Metric		System Property
• Ecological Footprint	↔	Ecological Burden
• Exergy	↔	Energy Efficiency
• Emergy	↔	Energy Resources
• Green Net Prod	↔	Economy
• Fisher Information	↔	System Order

3

EPA United States Environmental Protection Agency

P-GRAPH FRAMEWORK

EPA United States Environmental Protection Agency

P-Graph Process Synthesis (Friedler and Fan, 1992-)

Synthesis →

- Building blocks:
  - **materials:**
    - raw materials: ▼
    - products: ◎
    - intermediates: ●
  - **operating units:** —

- Process structure:

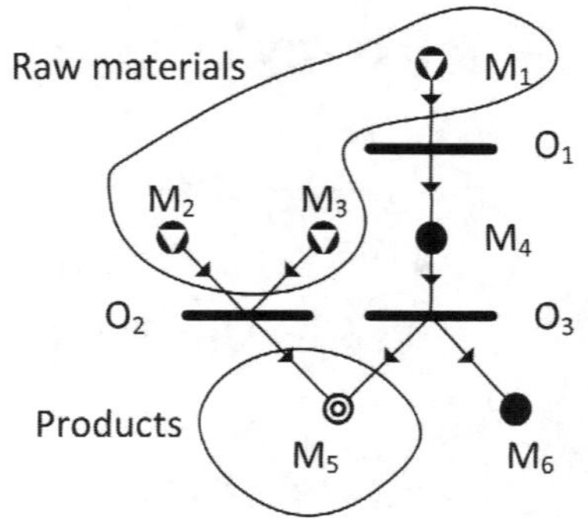

P-graph methodology: Structural Representation

- Clear logical interpretation
  - Each precondition (E **AND** F) has to be satisfied to operate an activity (O_2)
  - Any of the activities (O_2 **OR** O_3) having the same effect can potentially be sufficient to initiate the consecutive step (O_1)

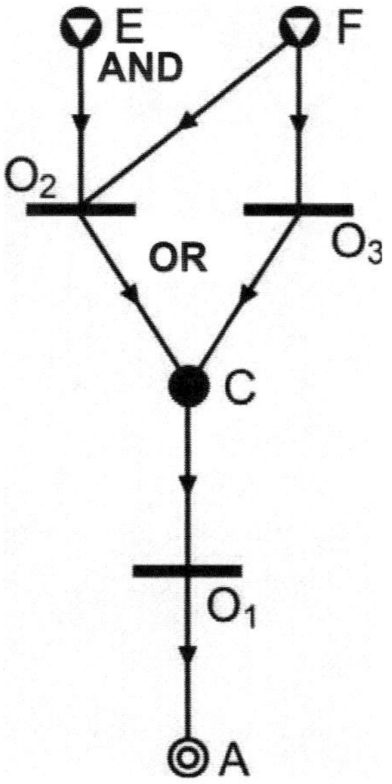

[redacted]

P-Graph Methodology

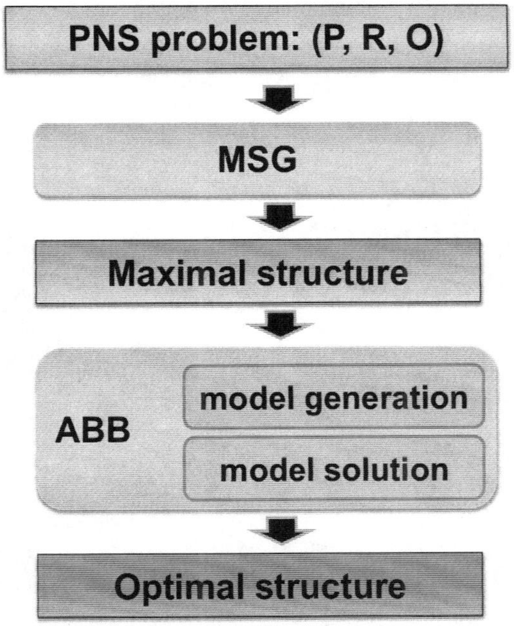

[redacted]

United States Environmental Protection Agency

SUSTAINABILITY METRICS

8

United States Environmental Protection Agency

Ecological Footprint Basics

- Land is categorized into:
  - –Arable land
  - –Forest land
  - –Pasture land
  - –Sea
  - –Energy land
  - –Built land

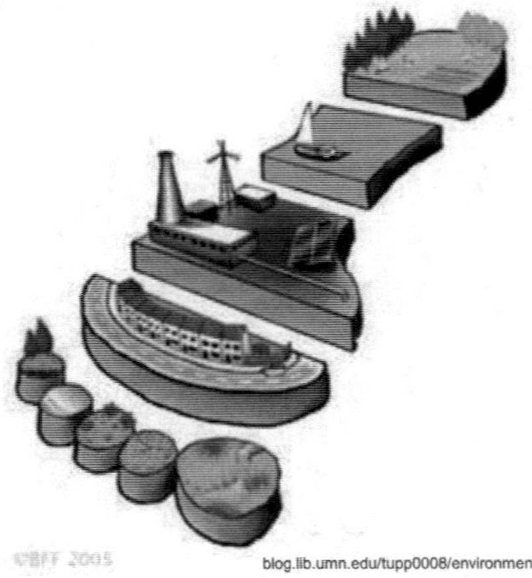

blog.lib.umn.edu/tupp0008/environment/2008/02/

9

Ecological Footprint Basics

- Ecological Footprint (demand) = land area required to meet level of consumption and waste generation by the human population
- Biocapacity (supply) = land area available to support the human population

–Assumptions:

- Can track resources and waste generated
- Resource and waste flows can be converted to land area

Sustainability Criteria: (1) Biocapacity is larger than Ecological Footprint, (2) Ecological Footprint does not increase and Biocapaicty does not decrease . $B \geq EF \ \& \ \Delta EF \leq 0 \ \& \ \Delta B \geq 0$

10

Emergy: Energy Resources

Emergy: the energy resources (in solar joules) invested by the environment in an operation or in a product

Emergy: the sum of all of the different kinds of energy previously used up, directly and indirectly, to make an item

Sustainability Criteria for Supply Chains:

(Total Emergy of Imputs) → Minimum

(Renewable Emergy Use) / (Total Emergy Use) → 1

11

United States Environmental Protection Agency

SUPPLY CHAIN STRUCTURES

12

United States Environmental Protection Agency

Assumptions

- The ecological footprint of the supply chain is dominated by the feedstocks and inputs
- The emergy footprint of the supply chain is also dominated by the feedstocks and inputs
- Methodology must be simple enough for wide application
- Supply chain is designed to produce 7.2 TJ/year of electricity and 18 TJ/year of heat.

13

United States Environmental Protection Agency

Maximal Supply Chain Structure

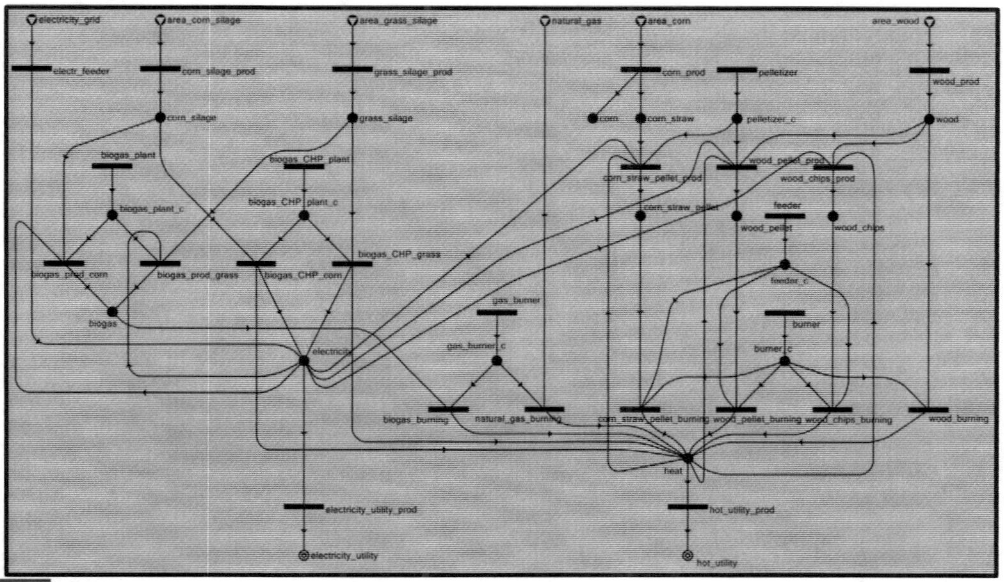

14

United States Environmental Protection Agency

Raw Material & Energy Inputs

Solution structures	Raw materials					
	electricity_grid [TJ/yr]	natural_gas [m³/yr]	area_corn [ha/yr]	area_corn_silage [ha/yr]	area_grass_silage [ha/yr]	area_wood [ha/yr]
Structure1	7.37					500.00
Structure2				117.69		367.73
Structure3					128.54	367.73
Structure4				120.03		393.66
Structure5	7.57					539.13
Structure6					131.10	393.66
Structure7		**399,272.00**		**116.30**		
Structure8			72.96	126.77		
Structure9				124.78		380.69
Structure10		**399,272.00**			**127.02**	
Structure11			72.96		138.46	
Structure12					136.29	380.69
Structure13	**7.25**	**540,588.00**				
Structure14	7.99					529.10
Structure15	8.17		102.04			
Structure16				**214.45**		
Structure17				125.12	98.01	
Structure18				90.05	135.88	
Structure19					234.67	
Structure20	7.95			125.00		
Structure21	8.02				136.36	

15

Supply Chain Structures

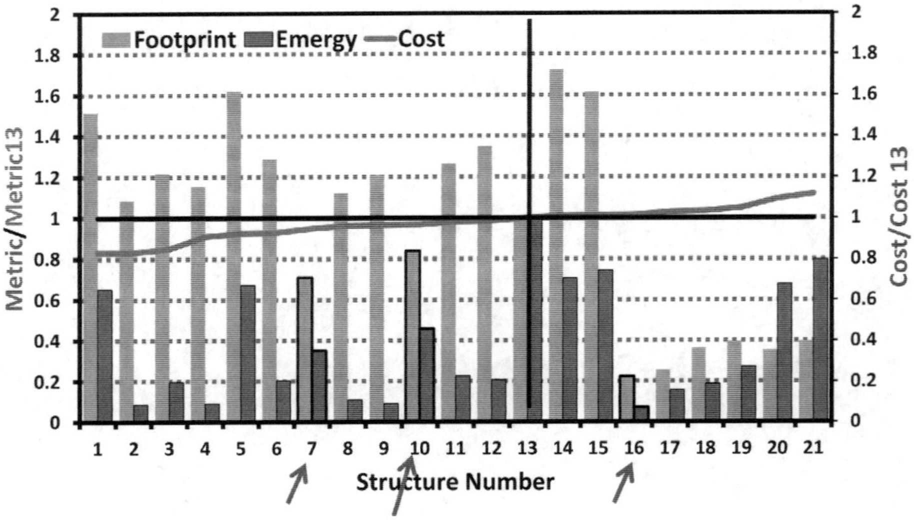

16

Structure #13: Natural Gas & Electricity from the "Grid"

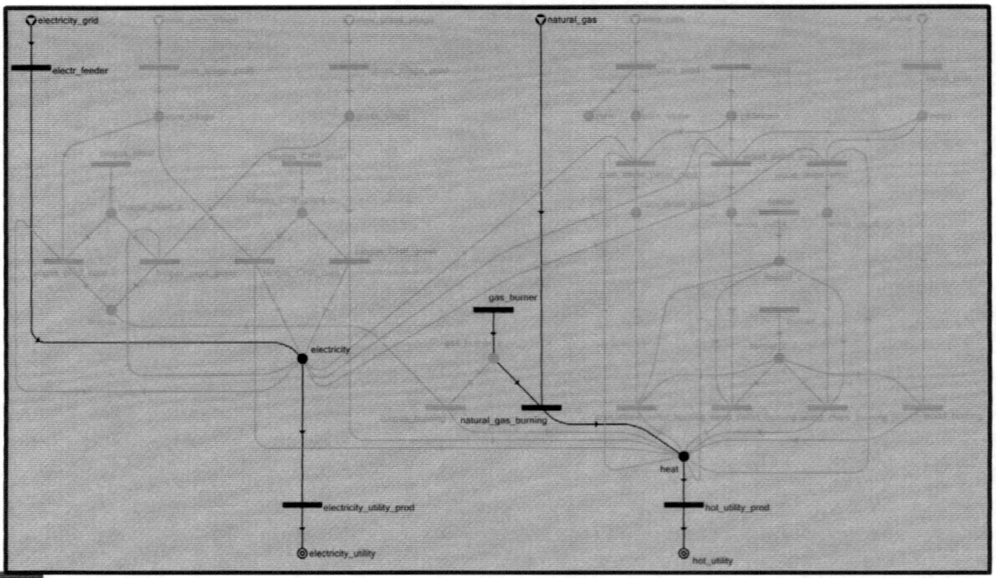

17

United States Environmental Protection Agency

Structure #7: Corn Silage & Natural Gas

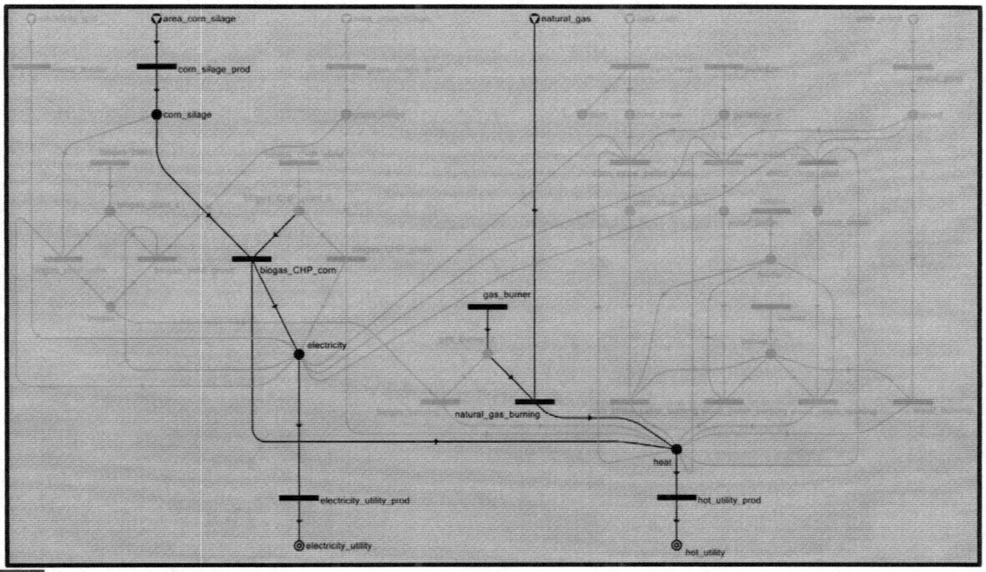

18

United States Environmental Protection Agency

Structure #10: Grass Silage & Natural Gas

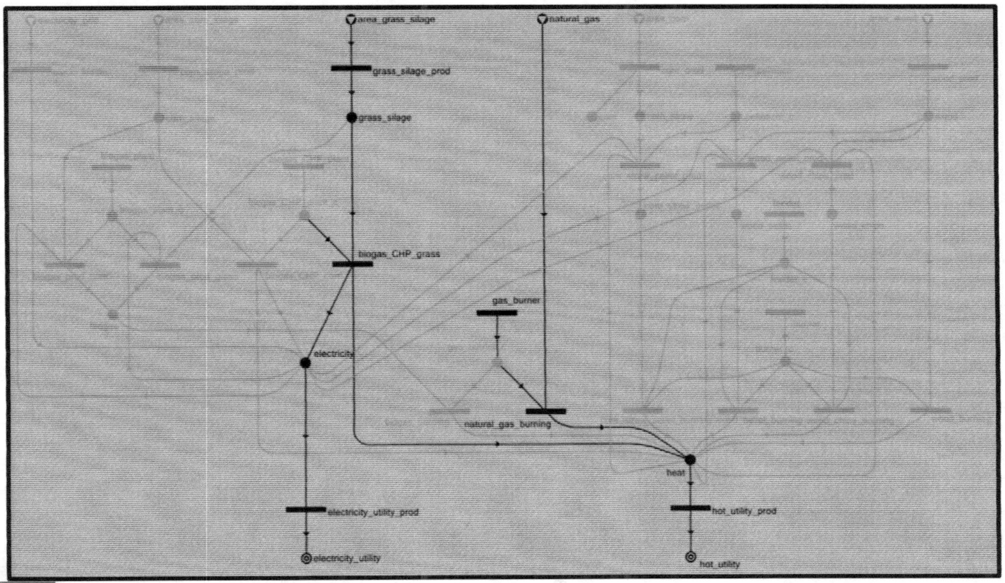

19

EPA United States Environmental Protection Agency

Structure #16: Corn Silage

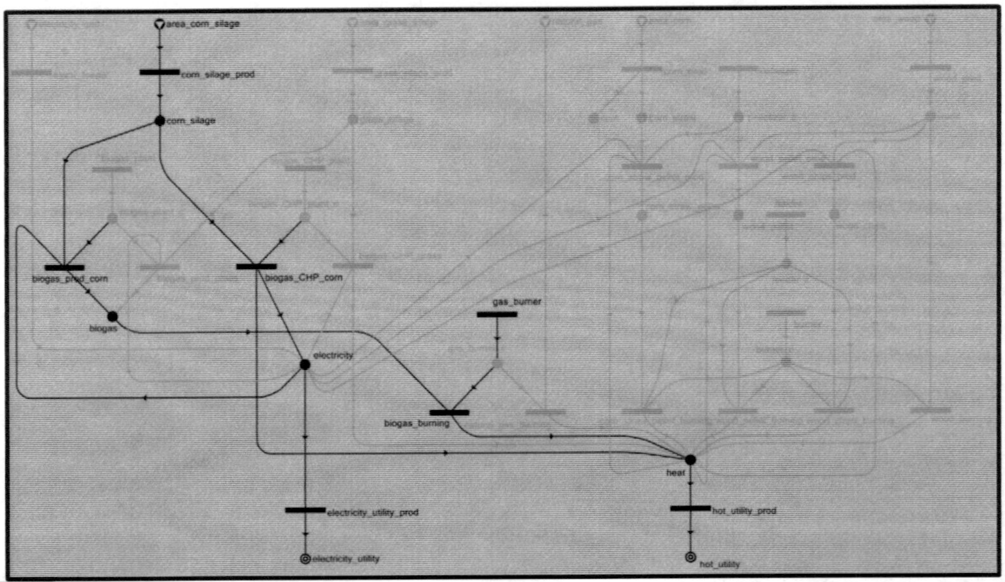

20

EPA United States Environmental Protection Agency

SCALE-UP

21

EPA
United States
Environmental Protection
Agency

Some Scale-Up Trade-Offs

More Heat & Electricity	Environmental Issue
• More natural gas and nuclear electricity	• Larger EF to process GHC emissions/nuclear waste; non-renewable emergy
• More agricultural waste	• Lower EF/more renewable emergy/more transportation
• More corn	• Lower EF/less food/more renewable emergy
• More wood fuel	• Lower EF/more transportation/ more renewable emergy

22

EPA
United States
Environmental Protection
Agency

SUMMARY

23

EPA United States Environmental Protection Agency

Summary

- Sustainability is about adaptively managing the environment on an on-going basis so as to insure that the Earth can continue to support human existence for the indefinite future.
- Supply chain design methodologies, e.g. p-graph framework, along with integrated sustainability metrics (emergy and ecological footprint) along with cost can provide insight into supply chain performance.
- Carefully designed supply chains can be made cheaper and more environmentally friendly than current practice while meeting societal needs.
- Delicate trade-off appear on scale-up of supply chain.

24

AIChE American Institute of Chemical Engineers

SECTION 5

Environmental, Health, and Safety Concerns

Process Design & Development | Process Safety |

Thinking Outside the Barrel for Biofuels

June 2013

Dr. Diana Matonis, PhD, MBA

Answer to the Ultimate Question of Life, the Universe, and Everything (42)*

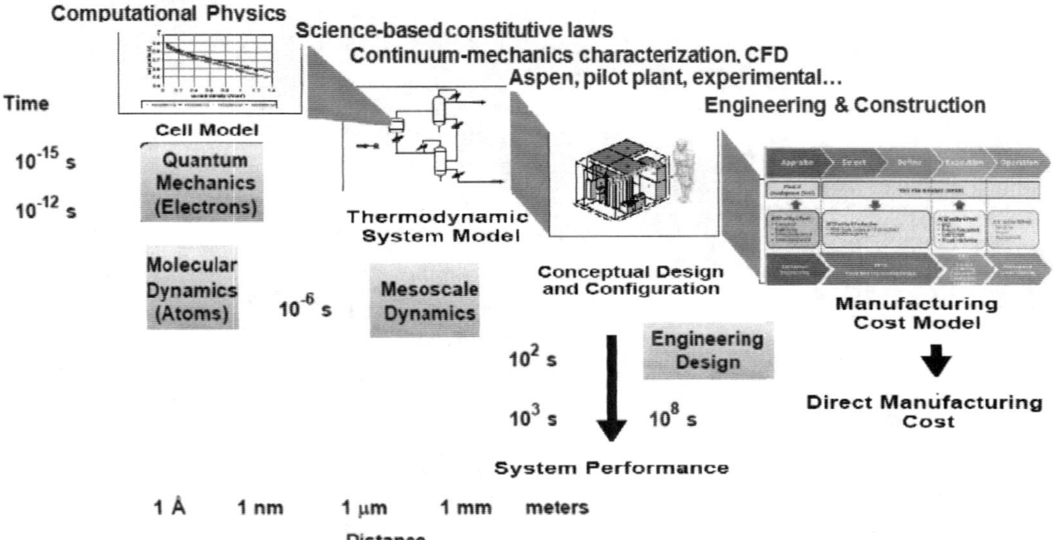

Source: D. Matonis, 2004

@2013, Flucomp Enterprises

* Quote from The Hitchhiker's Guide to the Galaxy

Thinking Outside the Barrel for Biofuels by Dr. Diana Matonis, Flucomp Enterprises, Inc. June 13, 2013

Today - STAGES OF ANALYSIS

INITIAL STAGE (IS)	IN R & D PROGRESS	PROCESS SIMULATORS (PS)	INDUSTRIAL APPLICATION (IA)
Literature review Networking Industrial innovators Process knowledge Industry Practices	CFD/Numerical model per Unit Operation Model development Model validation/bench mark Equipment testing or if new to Process Simulation	Demo/Pilot/Industrial application Process Parametric studies Solve problems Improvements Sizing/Reduce Capital Costs	Technology transfer and release

Initial Stage (IS) - Our Expectations of Timeline to Renewables

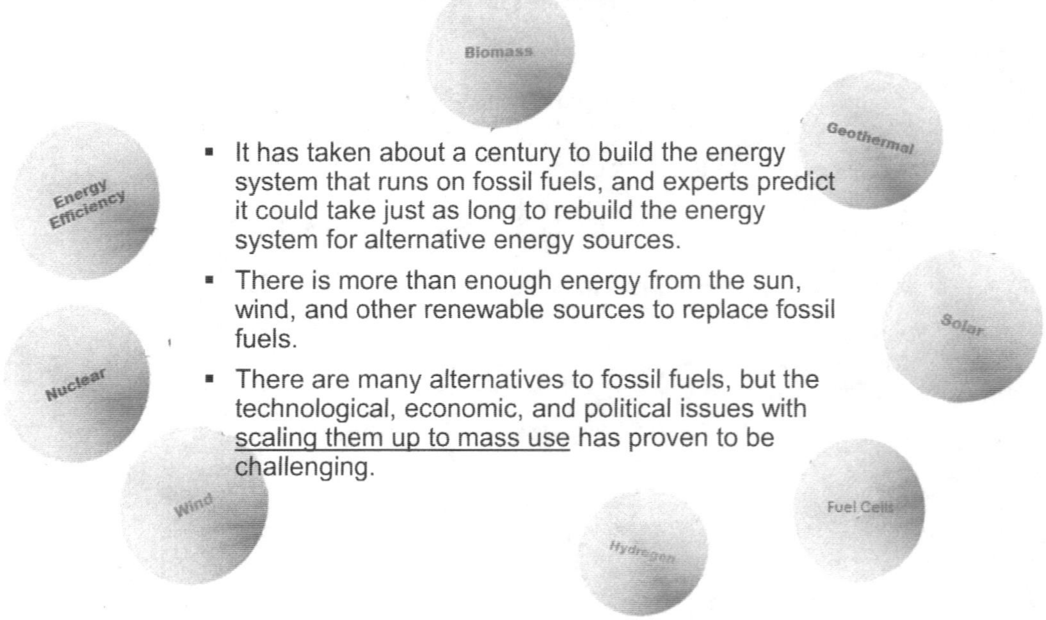

- It has taken about a century to build the energy system that runs on fossil fuels, and experts predict it could take just as long to rebuild the energy system for alternative energy sources.
- There is more than enough energy from the sun, wind, and other renewable sources to replace fossil fuels.
- There are many alternatives to fossil fuels, but the technological, economic, and political issues with scaling them up to mass use has proven to be challenging.

IS - Overview of Energy Full Fuel Cycles

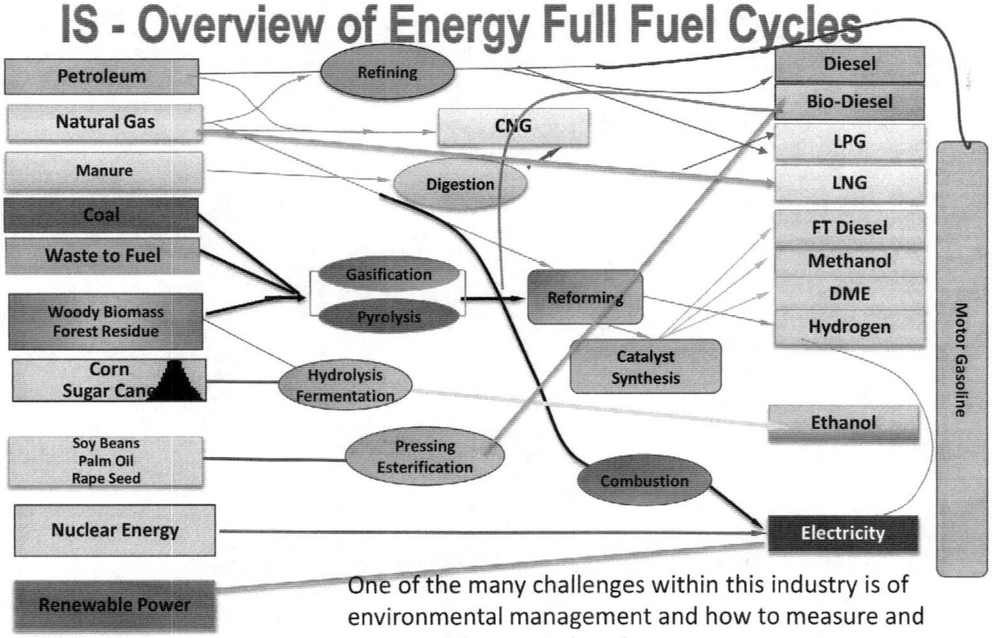

One of the many challenges within this industry is of environmental management and how to measure and assess environmental performance.

Source: Diana Matonis

IS - Drivers to Change

- Social, technological, and economic forces drove the momentous shift from wood to coal, and then to oil and gas.
- One factor clearly is the growing scarcity of existing fuels to meet population growth, urbanization and energy demand.
  - However, today natural gas could keep up with rising demand until (maybe) 2050.
- So **energy shortages will really not be pushing** transitions to alternative sources for at least another 10 to 20 years?
  - The consumer as producers will opt for price if option exists freely

IS - Carbon Dioxide Footprint by Power Generation Industry

The amount of CO_2 produced by renewables is significantly less than conventional fuels.

Carbon dioxide released for one kilowatt-hour produced (grams CO_2/kilowatt-hour) by major power generation methods

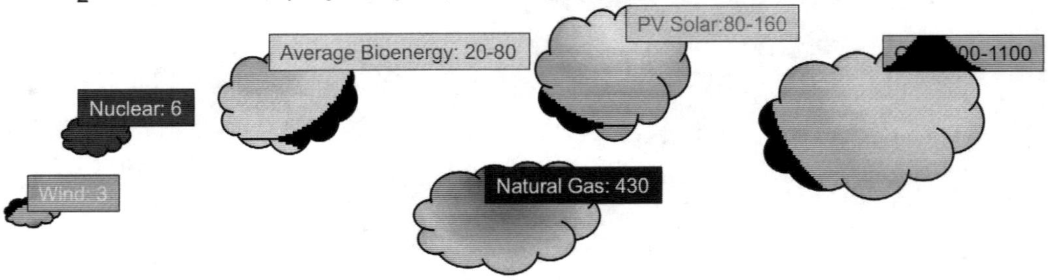

There is no single GHG solution, but with commitment, timing and correct investment decisions, business has the opportunity to make a difference

Source: The 'Carbon Inventory' of ADEME", May 2004

IS - Power Density the Face of Renewables Challenges

- Renewables are more challenging as an energy source for several reasons:
- Biofuel sources like corn or wood chips tend to be bulky an their energy content is diffuse.
  - A kilogram of oil has three times as much energy as a kilogram of biomass. The "power density" a coal mine or oil field produces 50 times more power per square meter than a solar facility, <u>10 to 100 times more than a wind farm</u>, and 100 to 1000 times more than a biomass plant.

Power Source	Power Density (W/m^2)	
	Low	High
Natural Gas	200	2000
Coal	100	1000
Solar (PV)	4	9
Solar (CSP)	4	10
Wind	0.5	1.5
Biomass	0.5	0.6

Source: MasterResource, A free-market energy blog

Thinking Outside the Barrel for Biofuels by Dr. Diana Matonis, Flucomp Enterprises, Inc. June 13, 2013

R&D - Municipal Solid Waste

- Given the scale of power density of biomass – I decided with a few others to use waste for fuels and patent and build.

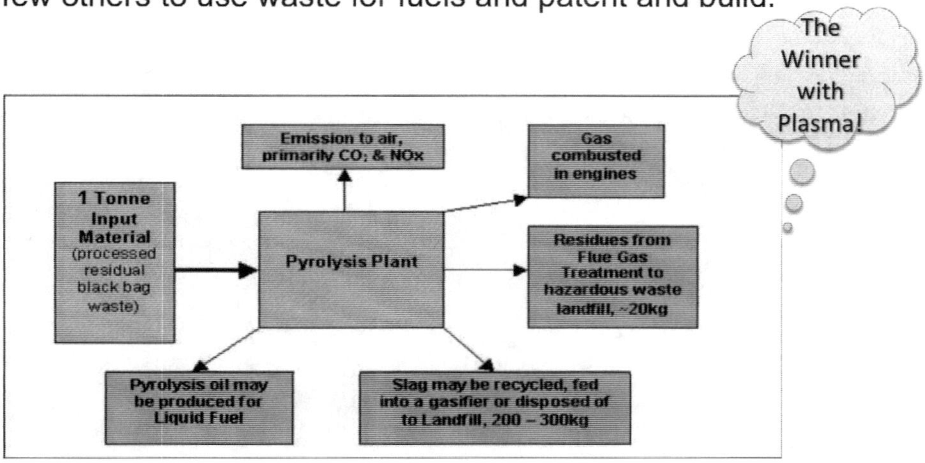

Thinking Outside the Barrel for Biofuels by Dr. Diana Matonis, Flucomp Enterprises, Inc. June 13, 2013

Waste Pyrolysis in Rotary Kilns

- Rotary kiln is a closed-system rotating cylinder - oxygen free atmosphere.
  - The pyrolysing energy comes over the outside heated kiln shell.
- Some very interesting effects from this process:
  - Low process temperature of ~900 - 1400 F with a residence time of upto 1 hour
  - Gas treatment due to the product input energy
  - Cracking of dioxines and furanes
  - Getting a product which can be sold
    - Metals pre-sortment/treatment
      - The ferrous and non-ferrous/sand/grit residues are separated by a magnetic separator and eddy current and fed to containers by belt feeders.
      - The aluminum residue (2%) goes to another company for further recycling.

Thinking Outside the Barrel for Biofuels by Dr. Diana Matonis, Flucomp Enterprises, Inc. June 13, 2013

Municipal Waste Feed Composition

Available Carbon from different MSW Sources

Component of waste stream	Carbon (% by Weight)	Oxygen (% by Weight)	Nitrogen (% by Weight)	Sulfur (% by Weight)	Ash (% by Weight)
Paper	43.5	6	44	0.3	0.2
Cardboard	44	5.9	44.6	0.3	0.2
Plastics	60	7.2	22.8	-	-
Textiles	55	6.6	31.2	4.6	0.2
Rubber&Leather	69	9	5.8	6	0.2
Wood	49.5	6	42.7	0.2	0.1
Glass	0.5	0.1	0.4	<0.1	-
Metals	4.5	0.6	4.3	<0.1	-
Other	26.3	3	2	0.5	0.2

	C6H12O4	Analysis		
	% of Biomass	Cellulose	Hemi-cellulose	Total
Corn Stover	38.40%	26.68	42.93	69.61
Yard waste	26.50%	18.41	12.80	31.22
Food waste	76.40%	53.09	0.00	53.09
Wood Scraps	45%	31.27	34.52	65.79
Corrugated Newspapers	57.30%	39.82	12.43	52.24
Newspaper	48.50%	33.70	11.30	45.00
Office paper	87.40%	60.73	10.54	71.28
Mixed Paper	42.30%	29.39	11.80	41.19

Thinking Outside the Barrel for Biofuels by Dr. Diana Matonis, Flucomp Enterprises, Inc. June 13, 2013

Kiln CFD Analysis

Input waste = 13,000 kg/h

Output = Pygas – quantity 10,300 kg/h, energy 55.3 MW

Pyrolysis char – quantity 2,500 kg/h, energy 8.3 MW

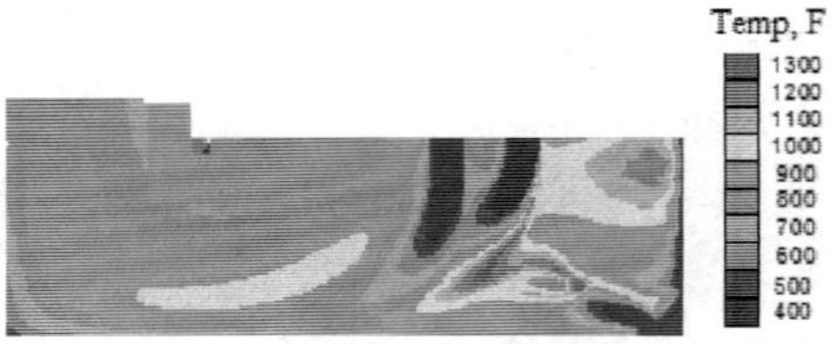

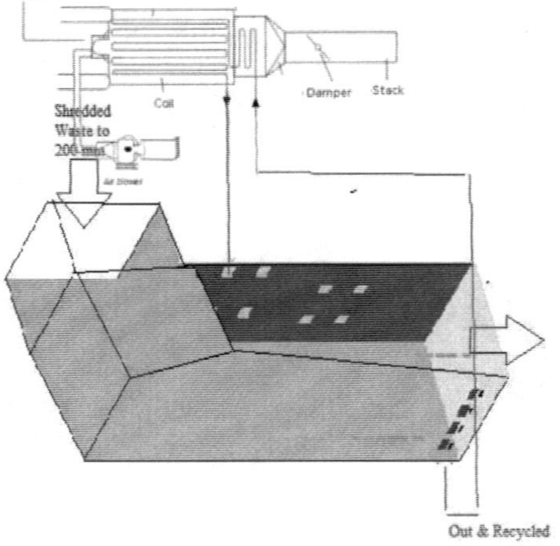

Actual Data Points

Pot Speed	Feed Rate	T (°F), zone 1	T (°F), zone 2
1.5	1.5	1076	902
1.5	1.75	1258	1058
1.5	1.75	1064	889
2	1.75	1284	1059
2	2	1274	1016
2	2	1302	1178

Kiln Exit gases Analysis By weight (By Orsat Gas Analyzer)			
Mass	Percent CO2	28.1	%
Mass	Percent N2	70.7	%

Thinking Outside the Barrel for Biofuels by Dr. Diana Matonis, Flucomp Enterprises, Inc. June 13, 2013

IS - Burning biomass Can Emit more CO_2 than fossil fuels per MW energy Generated

- Wood inherently emits more carbon per Btu than other fuels
  - Natural gas: 117.8 lb CO_2/mmbtu
  - Bituminous coal: 205.3 lb CO_2/mmbtu
  - Wood: 213 lb CO_2/mmbtu (bone dry) Wood is often wet and dirty, which degrades heating value
  - Typical moisture content of wood is 45 – 50%,
    - Before "useful" energy can be derived from burning wood, some of the wood's btu's are required to evaporate all that water.
- Biomass boilers operate less efficiently than fossil fuel boilers
  (Source: data from air plant permit reviews and the EIA)
  - Utility-scale biomass boiler: 24%
  - Average efficiency US coal fleet: 33%
  - Average gas plant: 43%

Thinking Outside the Barrel for Biofuels by Dr. Diana Matonis, Flucomp Enterprises, Inc. June 13, 2013

IS + R&D: What To Do with CO_2 In Addition to CCS?

- Produce clean syngas where the H2/CO ratio in a wide range of about 1.0 to 3.2?
  - Typical Fischer-Tropsch Process - H2/CO ratio at about 2.0
  - Typical syngas-to-SNG Process: H2/CO ratio over 3.0

    $CO + 3H_2 \rightarrow CH_4 + H_2O$

    $CO_2 + 4\ H_2 \rightarrow CH_4 + H_2O$

    - Lurgi, Haldor Topsoe and others license SNG technologies today
  - Plasma-assisted CO_2 dissociation offers one possible solution to reduce concentrations of the greenhouse gas.
    - CO and oxygen main products of the plasma-chemical reactions in MSW applications.
      - Dielectric barrier discharges
      - Microwave discharges
      - Glow discharges
    - NETL and many Companies (installed) have looked at Plasma Gasification of MSW Directly
      - http://www.netl.doe.gov/technologies/coalpower/gasification/gasifipedia/4-gasifiers/4-1-4-1a_westinghouse.html

Thinking Outside the Barrel for Biofuels by Dr. Diana Matonis, Flucomp Enterprises, Inc. June 13, 2013

IS & R&D Shows Carbon Emissions from Burning Biomass & Public Buy-In Issue

The News – What the public is being told that biomass is a “low carbon” or “carbon neutral” fuel or in layman terms => the carbon emitted by biomass burning won’t contribute to climate change

The Experience - What the public hear’s or experiences on large-scale installations & implementation

- Biomass burning power plants emit 150% the CO2 of coal, and 300 – 400% the CO_2 of natural gas, per unit energy produced.
- Example in the News:
  - Biomass facility at the Domtar paper Mill in Rothscild, WI
    - **Upgrade**: Install a new natural gas boiler alongside a new biomass boiler
    - **Permit**: Biomass boiler would emit 6 times more carbon (at 3,120 lb/MWh) than The adjacent natural gas turbine (at 510 lb/MWh)

Thinking Outside the Barrel for Biofuels by Dr. Diana Matonis, Flucomp Enterprises, Inc. June 13, 2013

IS - Renewable and Alternative for Transportation Fuel has Taken Off!

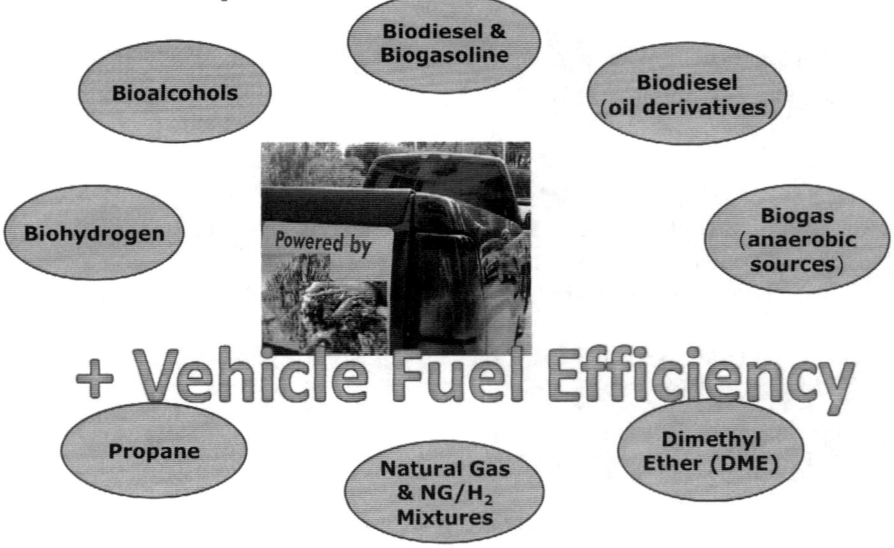

Thinking Outside the Barrel for Biofuels by Dr. Diana Matonis, Flucomp Enterprises, Inc. June 13, 2013

IS- Take a Look at Transportation Driver

- Today - Contributory Drivers:
  - Government incentives and mandates
  - Desire for less dependence on imported oil
  - Effort to increase farm income
  - Create jobs in rural communities
  - Environment will now continues to be the highest driver

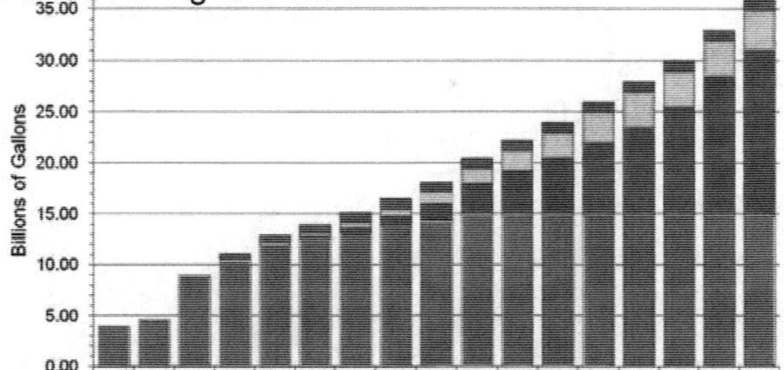

	2006	2007	2008	2009	2010	2011	2012	2013	2014	2015	2016	2017	2018	2019	2020	2021	2022
Biomass-based Diesel				0.50	0.65	0.80	1.00	1.00	1.00	1.00	1.00	1.00	1.00	1.00	1.00	1.00	1.00
Non-celulosic Advanced				0.10	0.20	0.30	0.50	0.75	1.00	1.50	2.00	2.50	3.00	3.50	3.50	3.50	4.00
Celulosic Advanced					0.10	0.25	0.50	1.00	1.75	3.00	4.25	5.50	7.00	8.50	10.5	13.5	16.0
1st Generation Growth of Corn Based Ethanol	4.00	4.70	9.00	10.5	12.0	12.6	13.2	13.8	14.4	15.0	15.0	15.0	15.0	15.0	15.0	15.0	15.0

Even As VMT and MPG May Rise, but GHG is Not

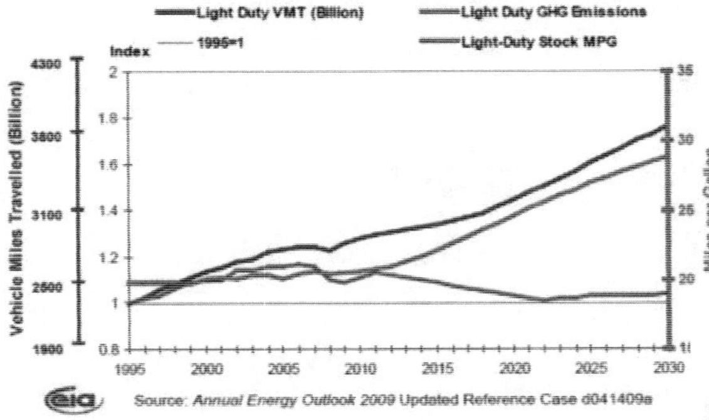

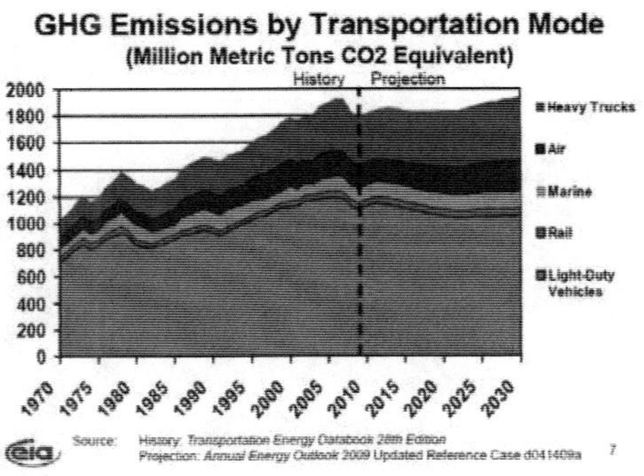

Thinking Outside the Barrel for Biofuels by Dr. Diana Matonis, Flucomp Enterprises, Inc. June 13, 2013

Challenges to Continue to Address

- **New Technologies**
  - The proposed engineered biomass process of Plasma can reduce CO_2 footprint, but more R&D and Actual Installation information and work is ongoing with demonstration scale plants
  - When consumers face truths about certain technologies not being so green, buy in diminishes
  - Integrate environment with design

- We ARE getting there on the large-scale
  - Non-sustainable technologies are removed from the pool of possibility by the laws of economics, not through prejudicial choice!

- Infrastructure, additional stressors
  - Large-scale implementation
    - Flex-Fuel Vehicles/Plug-In Hybrids
    - Decreases Justification of Petroleum for Transportation Needs

- Implementation Assistance continues in the form of
  - Incentives
  - New Policies

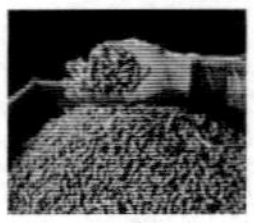

- Environmental Planning and Profiling
  - Establish well-defined but fluid impact monitoring programs
  - Integrate design with environment.
    - It is being done in the Oil & Gas Industry.
    - Maintain an open cross industry dialogue

Thinking Outside the Barrel for Biofuels by Dr. Diana Matonis, Flucomp Enterprises, Inc. June 13, 2013

The End!

Biomass can be a part of the Clean Energy and Cimate Solution, but There Is No "One Shoe Fits All Solution"

Thinking Outside the Barrel for Biofuels by Dr. Diana Matonis, Flucomp Enterprises, Inc. June 13, 2013

Changing the Way We Use Wood – Lessons Learned in Biorefinery Pilot Trials

Hemant Pendse

Professor and Chair, Chemical and Biological Engineering Dept.
Director, Forest Bioproducts Research Institute (FBRI)
University of Maine, Orono, ME

2013 AIChE Process Development Symposium
Process Scale-Up and its Challenges

June 11-13, 2013
Oak Brook, IL

Outline

- Motivation – Forest Biorefinery
  - Lignocellulosic conversion to fuels, chemicals & advanced materials
  - Value chain for wood utilization
  - Pre-pulping extraction of hemicellulose
  - Moving through technology readiness levels
- Environmental Benefits vs Environmental Assessment
  - recovering hemicelluloses that would otherwise be wasted
  - off-loading lime kiln fossil fuel demand
  - potential environmental impacts associated with repurposing pulp mill
- Wood-derived sugar opportunities
  - Mixed C6 & C5 sugars
  - Cellulosic sugars from certified sustainable and renewable feedstock
  - Satellite biorefinery options

MAINE

Bringing Research to Marketplace through Rapid Commercialization

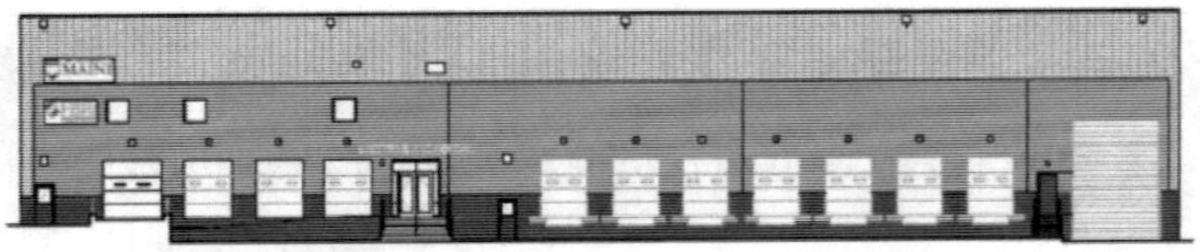

Fuels

Chemicals

Advanced Materials

MAINE

FBRI Philosophy

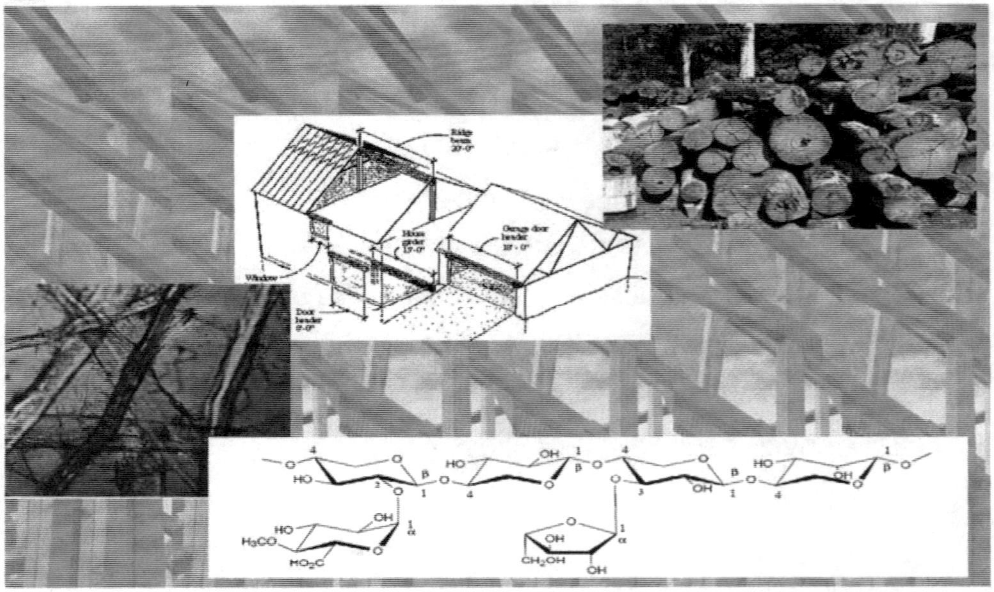

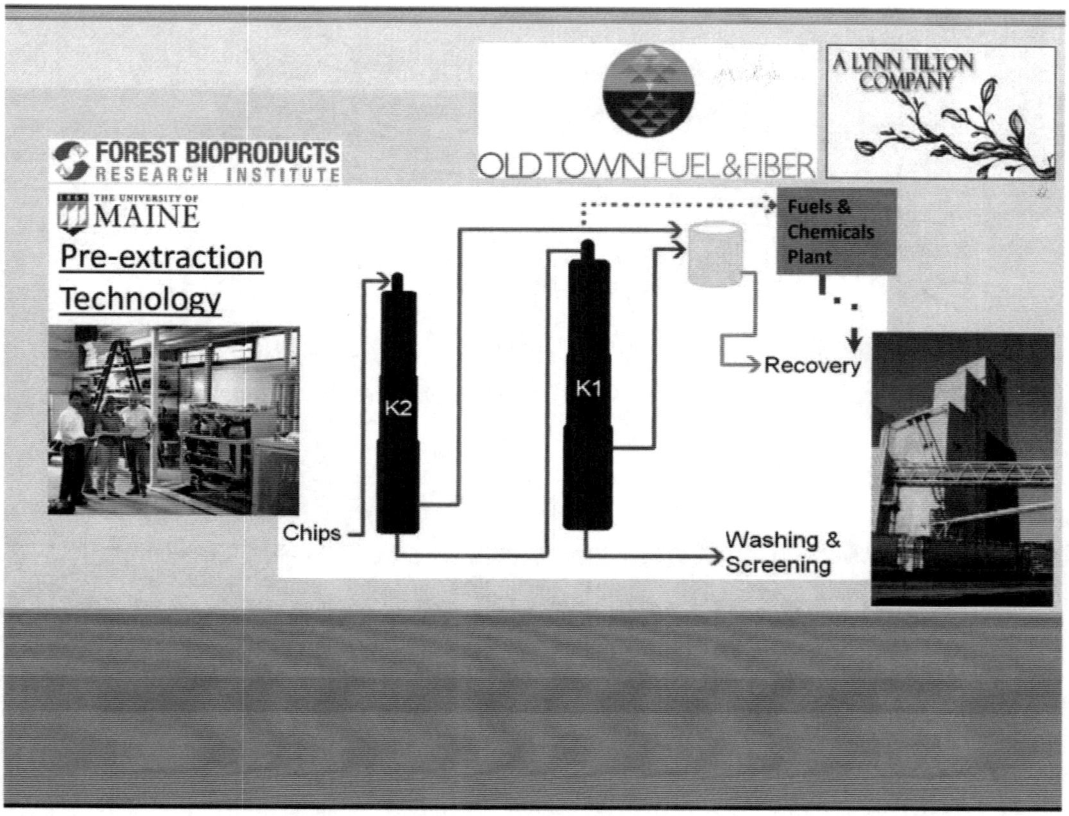

MAINE

DOE, NEPA, and You

A Guide to Public Participation

How Does NEPA Work?

Early in its planning process for a proposed action, DOE considers how to comply with the National Environmental Policy Act (NEPA). The appropriate level of review depends on the significance, that is, consideration of the context and intensity of the potential environmental impacts associated with the proposed action.

- Land use
- Water resources
- Biological resources
- Air quality
- Aesthetics
- Safety and occupational health
- Waste management and hazardous materials
- Utilities
- Cultural resources
- Traffic
- Socioeconomics and environmental justice

- Specific areas of onsite soils and noise analysis were excluded because of anticipated no impact.

R&D Trials & Campaigns and Potential To Emit (PTE) Analysis

- VOC, GHG and other criteria pollutants are in focus
  - Mill trials and relevant monitoring
  - FBRI TRC batch campaigns and relevant estimation

In Maine, any facility with air pollutant emissions that is not regulated as a major source is automatically designated as a minor source unless they can show their emissions are exempt from permitting. Chapter 115 provides a listing of the exemptions that can apply. The only exemption threshold that FBRI may exceed for minor source air permitting is listed at Section 1.C.(2)(d) which exempts any facility from air permitting if their normal operational process emissions of any regulated pollutant are less than 100 pounds per day or 10 pounds per hour.

Technology Research Center Scope

- **Innovation**
  TRL 1 – Basic Research
  TRL 2 – Applied Research
  TRL 3 – Critical Function or Proof of Concept Established
- **Emerging Technologies**
  TRL 4 – Laboratory Testing/Validation of Component(s)/Process(es)
  TRL 5 – Laboratory Testing of Integrated/Semi-Integrated System
  TRL 6 – Prototype System Verified
- **Systems Integration**
  TRL 7 – Integrated Pilot System Demonstrated
  TRL 8 – System Incorporated in Commercial Design
- **Market Penetration**
  TRL 9 – System Proven and Ready for Full Commercial Deployment

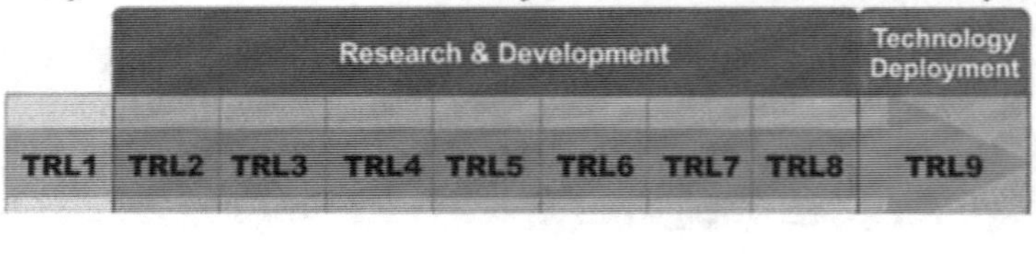

Bench Scale Extractor

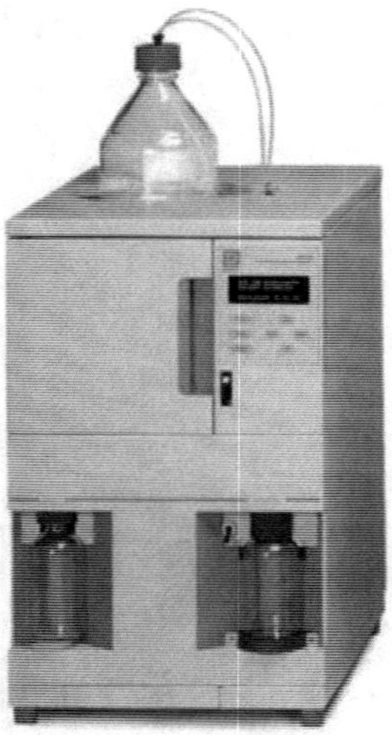

- Automated extraction of a single sample
- Economically priced for lower throughput laboratories
- Sample cell sizes: 10, 34, 66, and 100 mL
- Collection bottle 250 mL
- Operating pressure 1500 PSI (100 Bar)
- Small footprint requires less than 36 cm (14 in.) of bench space

Floor-scale extraction Unit

- Four digesters in rotating frame
- Digester volume: 60 liter /each
- Max. pressure: 195 psi
- Max. temperature: 205°C
- P + T wireless monitoring by Honeywell XYR 5000 transducers
- Honeywell Experion distributed control system
- Simultaneous, independent computer control of digesters

Floor-scale Extraction Unit

Floor-scale Extraction Unit – Safety Considerations

Mill trials

Up to 2000 tons per day woodchips feed to the pre-pulping extraction vessel (90 ft tall)

Key parameters:
- Liquor temperature
- Liquor pH
- Liquor to wood ratio
- Liquor recycle fraction
- Wood chip residence time

Mill trials

Challenges:

- Lignin fate
- Extract collection
- Handling partially-cooked woodchips after extraction
- Extract recycle and reuse

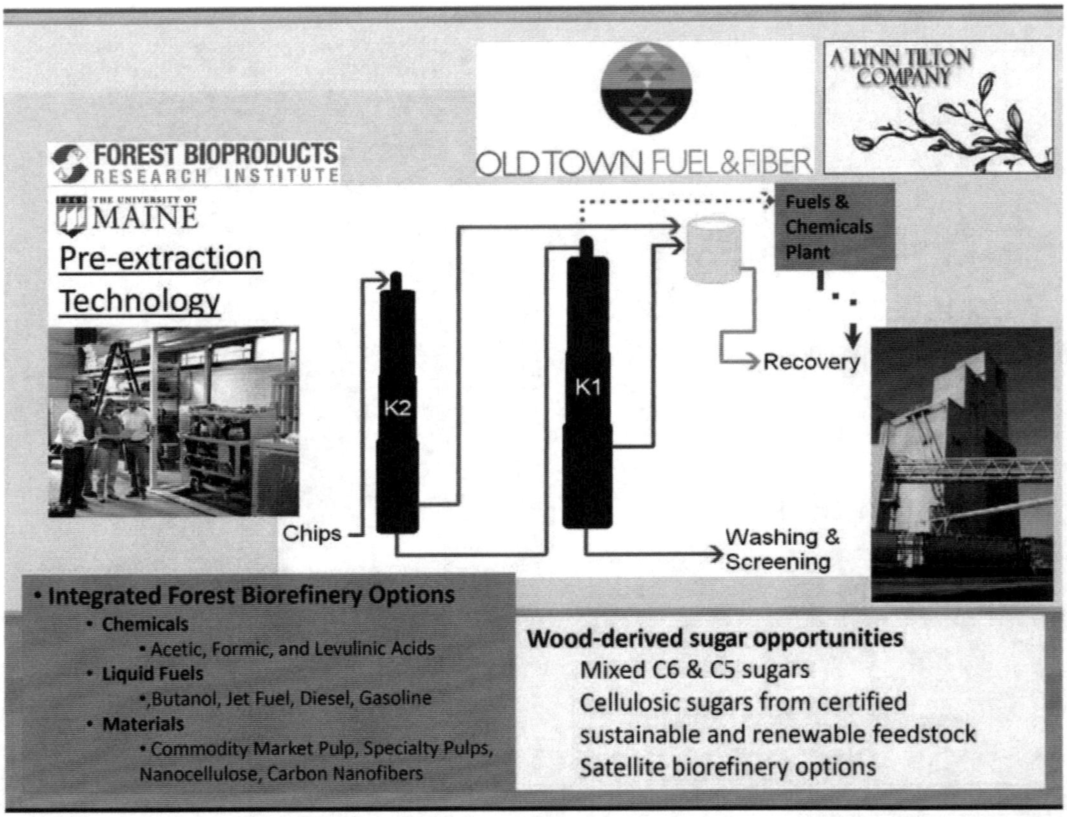

MAINE

Acknowledgements

- Support from OTFF project “Integrated Forest Products Biorefinery in Old Town, ME” on DE-EE0003364
- Special thanks to Amy Luce, Bill Ceckler, Joe Genco, Adriaan van Heiningen, and OTFF staff

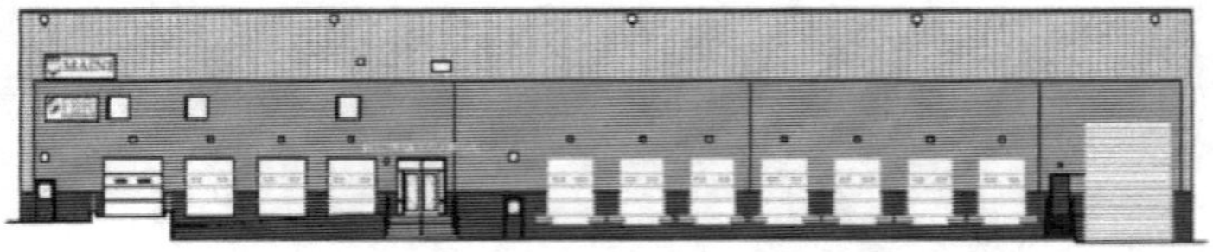

Hydrodynamics of FCC Risers: Advanced Simulation Tools

Reza Mostofi and Lev Davydov
UOP LLC, A Honeywell Company

2013 AIChE Process Development Symposium
June 11 - 13, 2013
Oak Brook, Illinois

UOP
A Honeywell Company

© 2013 UOP LLC. All rights reserved.

UOP 5978-1

Outline

UOP
A Honeywell Company

- UOP Business Profile
- Fluid Catalytic Cracking (FCC) Riser
- Computational Fluid Dynamics (CFD)
- Typical CFD Workflow
- Gas-Solids CFD Approaches
- Modeling FCC Riser
- CFD Results
- Summary

UOP 5978-2

UOP Business Profile

UOP
A Honeywell Company

- **Established in 1914**
- **Process Technology for Petroleum Refining, Petrochemical and Gas Processing industries**
- **Catalyst and adsorbent products for petrochemical and petroleum refiners**
- **Molecular sieves and alumina products for the manufacturing, chemical processing and natural gas industries**
- **Research & Development**
- **Engineering, Services, and Equipment**
- **Renewables Technology**

Innovative Technologies for Sustainable Growth & Customer Satisfaction

UOP 5978-3

FCC Riser

UOP
A Honeywell Company

- **Cracking reaction in riser**
- **Contact of hot catalyst with feed**
- **Tend to establish a core-annular structure:**
  - Vapor at center
  - Catalyst at wall
- **Desired to approach plug flow reactor**
- **Modeling the riser can help understand and improve the flow behavior and result in increased production of desired products**

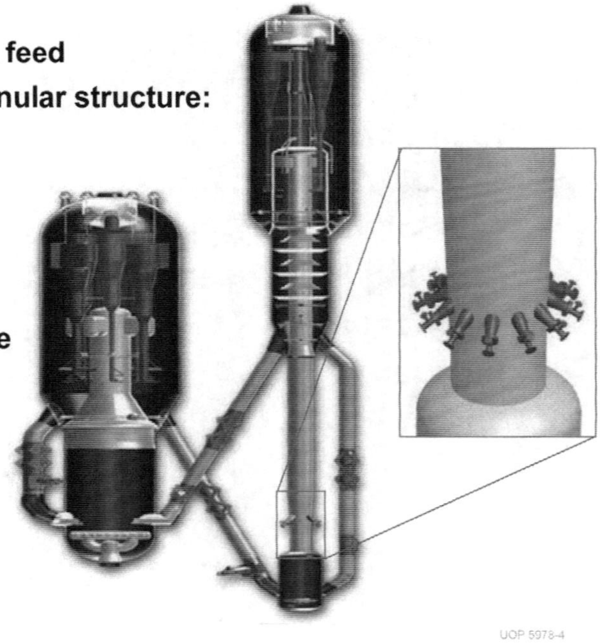

UOP 5978-4

Why Riser Hydrodynamics are Important

UOP A Honeywell Company

	Real riser		Close to ideal riser	Hydrodynamic effect on yields
Solids Holdup				Dense annulus can lead to overcracking at walls
Vapor Velocity				Strongly parabolic (triangular) profile can lead to conversion loss in core
Residence Time Distribution				Broad res. time dist. can lead to inferior yield patterns

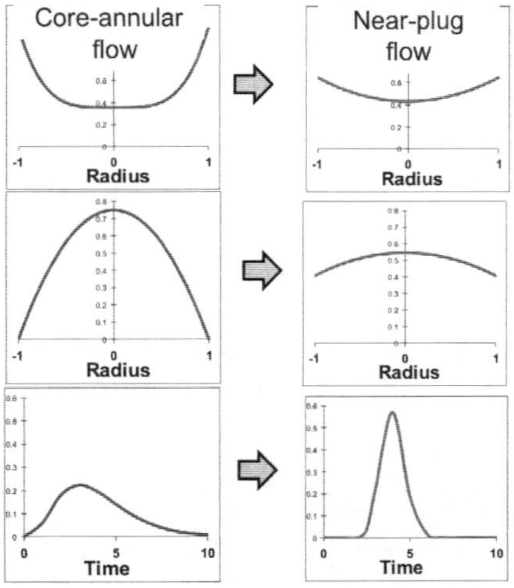

Larger diameter riser → higher slip factor → stronger hydrodynamic effects

UOP 5978-5

Computational Fluid Dynamics (CFD)

UOP A Honeywell Company

- **Computational Fluid Dynamics is numerical solution of the fundamental laws of physics combined with empirical correlations applied to a defined space at any time.**
  - ∅ Best suited for predicting trends rather than exact data
  - ∅ Lack of well established models to describe some aspects of multiphase processes
  - ∅ Including all process features in a CFD model sometimes means huge number of cells and hence lengthy runtimes for existing computer hardware

  - ✓ Fast, easy, more flexible and inexpensive compared to physical experiment
  - ✓ Provides detailed information about the elements in a system
  - ✓ Helps to determine impact of changes to a system

UOP 5978-6

Typical CFD Workflow

UOP
A Honeywell Company

Create Geometry

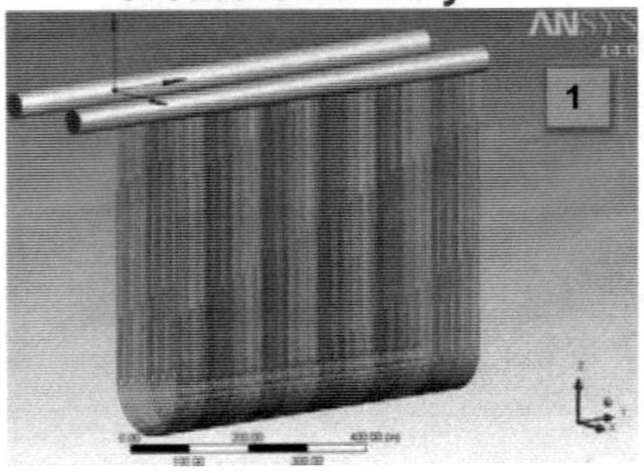

Generate CFD Mesh

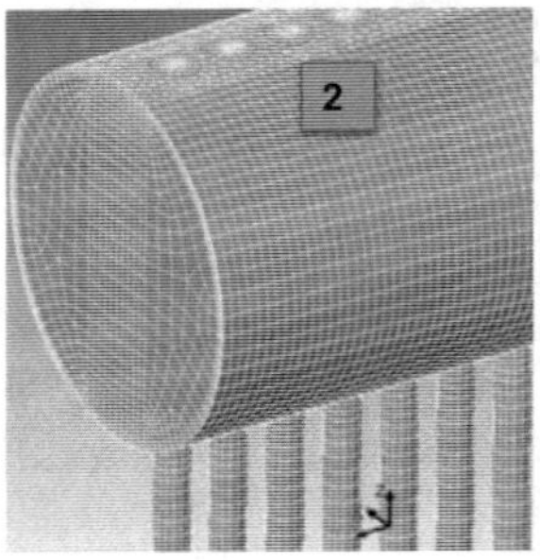

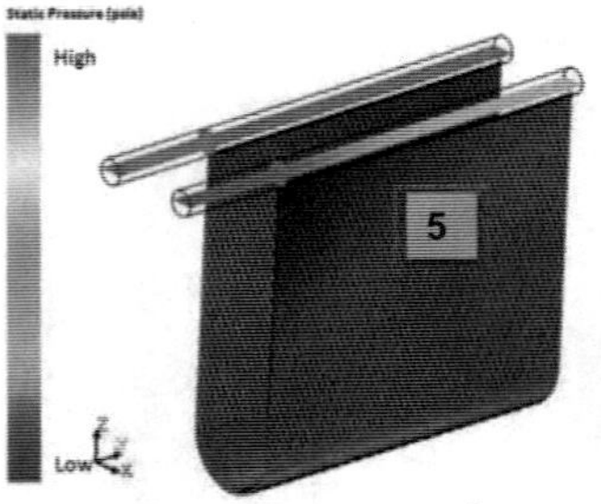

Post-Process & Analyze Results

Solve

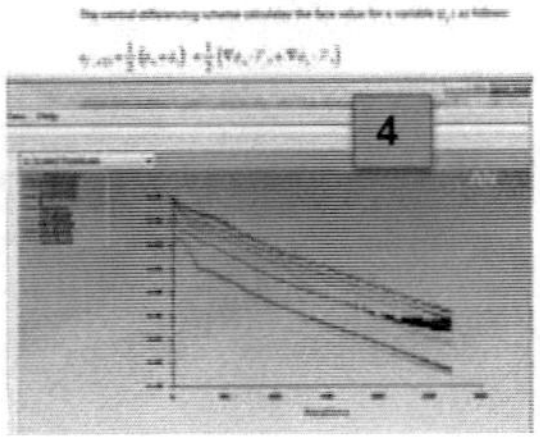

Set up Models & Optional Coding

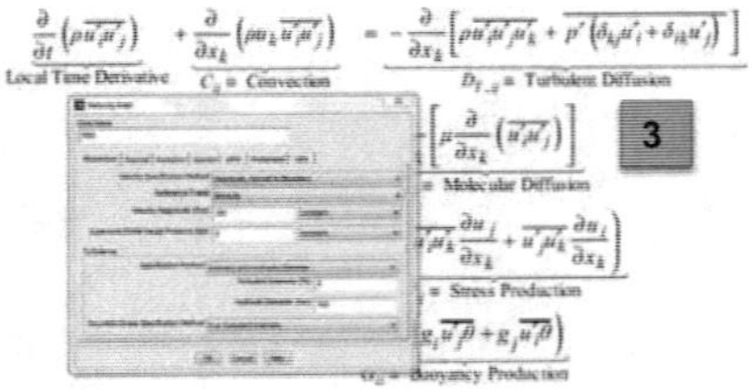

UOP 5978-7

Gas-Solids CFD Approaches

UOP
A Honeywell Company

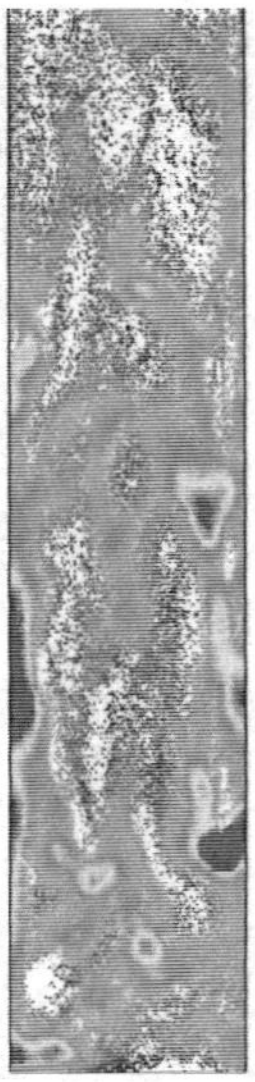

Particles Size using DPDPM model

- **Multiphase models are generally divided in three categories:**
  - Eulerian; the particles are treated as fluids (TFM)
  - Langrangian; individual particles are tracked (DEM)
  - Hybrid; combination of above models is used (DPDPM)
- **Each approach has pros and cons**
  - TFM is fast but less accurate when dealing with dense systems
  - DEM is very slow and almost impractical for industrial particle-laden flows
  - Hybrid models are still developing and do not cover all range of flow conditions

Catalyst Density Profile with TFM

UOP 5978-8

Modeling FCC Riser

UOP
A Honeywell Company

- **Both TFM and DPDPM approaches are presented for two flow conditions**
- **Expect to see different behaviors for these cases based on literature**

	Gas Superficial Velocity	*Catalyst Flux*	*Air Density*	*Particle Size* (1)	*Particles Density*
Condition	fps	$lb/ft^2/sec$	lb/ft^3	Microns	lb/ft^3
A	22.9	149.9	0.12	80	90
B	50.7	250.4	0.10	80	90

(1) For dpdpm there is a particle size distribution

Gas/Solids Outlet

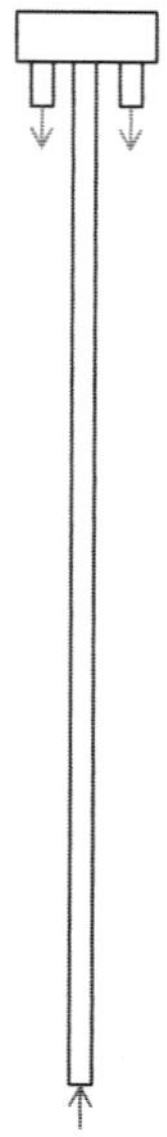

Gas/Solids Inlet

UOP 5978-9

Case A-TFM & DPDPM-Solids Flux

UOP
A Honeywell Company

CFD Predictions
U=23 fps, G=150 #/ft2s

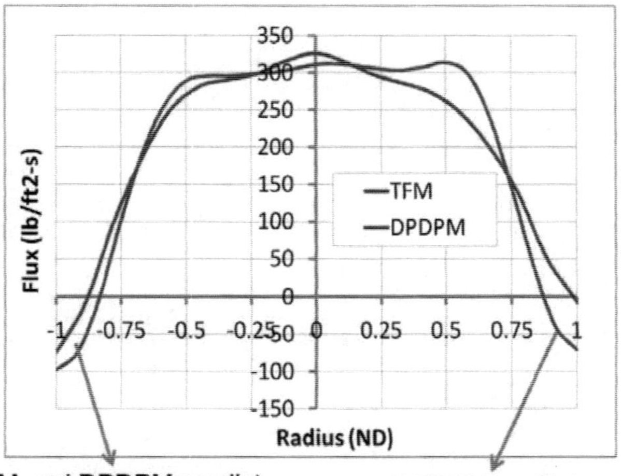

TFM and DPDPM predict downflow at wall

DPDPM predicts thicker annulus

Knowlton et al, Hydrodynamic Scale-Up of Circulating Fluidized Beds, *Fluidization XII* (2007)

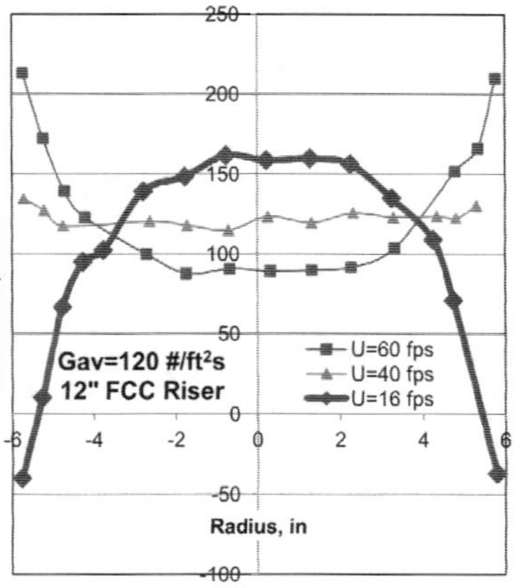

UOP 5978-10

Case A-TFM & DPDPM-Solids Density

UOP
A Honeywell Company

CFD Predictions
U=23 fps, G=150 #/ft2s

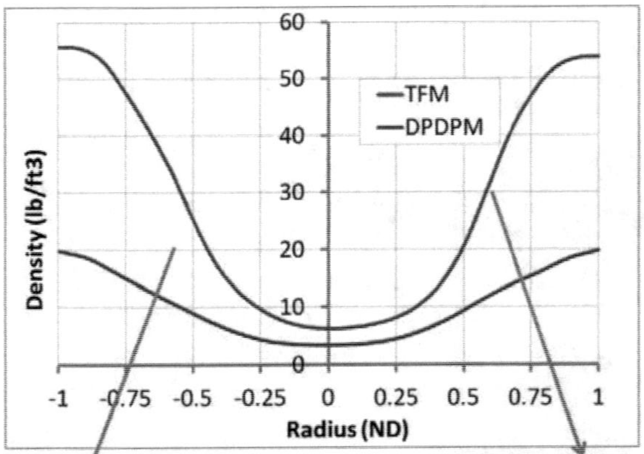

Knowlton et al, Hydrodynamic Scale-Up of Circulating Fluidized Beds, *Fluidization XII* (2007)

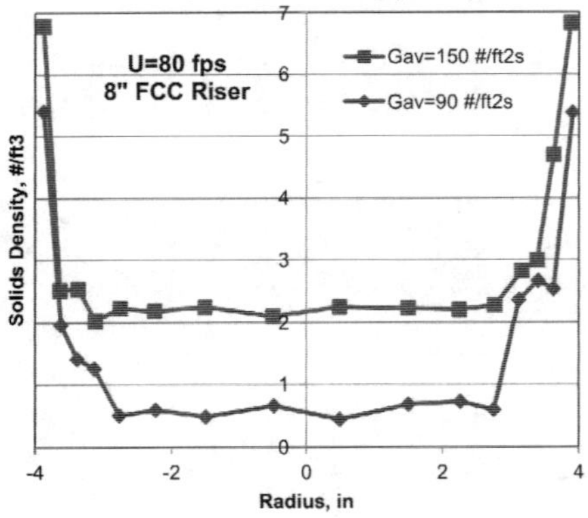

Core-annular flow structure

DPDPM predicts more pronounced core-annular structure

UOP 5978-11

Case B-TFM & DPDPM-Solids Flux

UOP
A Honeywell Company

CFD Predictions
U=51 fps, G=250 #/ft2s

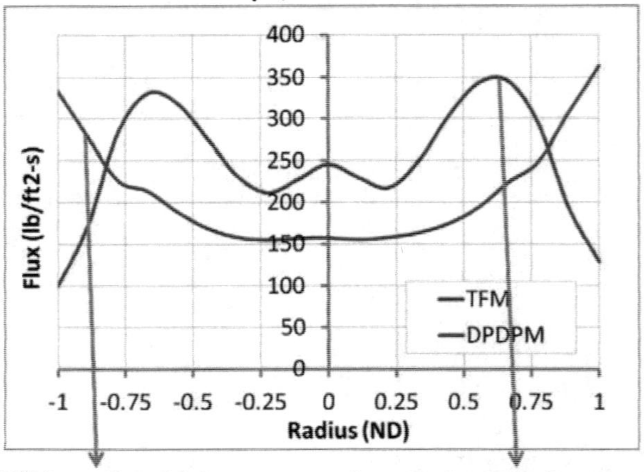

Knowlton et al, Hydrodynamic Scale-Up of Circulating Fluidized Beds, *Fluidization XII* (2007)

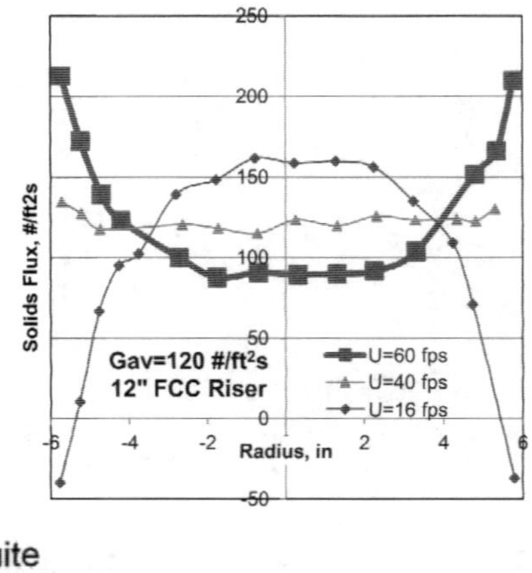

DPDPM predicts higher flux at the wall

Inverted parabola not quite established with TFM

UOP 5978-12

Case B-TFM & DPDPM-Solids Density

A Honeywell Company

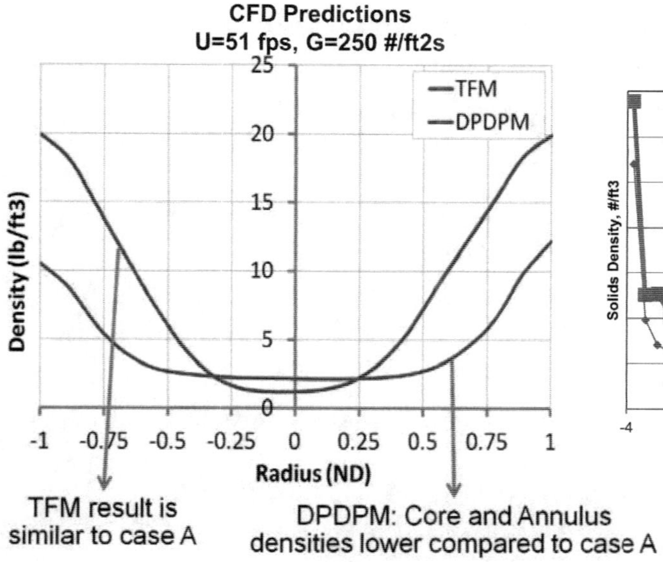

Knowlton et al, Hydrodynamic Scale-Up of Circulating Fluidized Beds, *Fluidization XII* (2007)

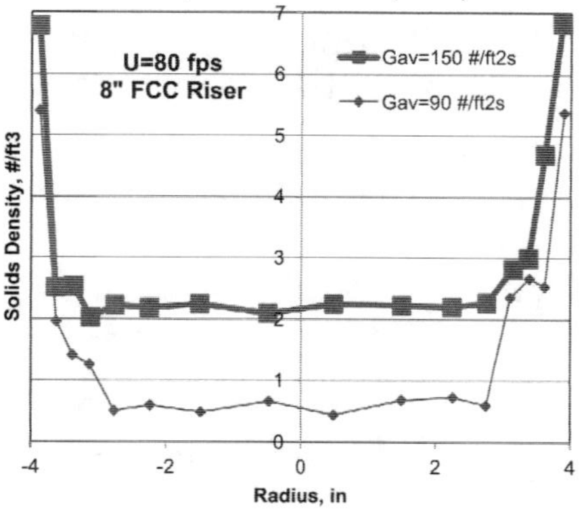

UOP 5978-13

Riser Diameter Effect – Velocity Profiles (CFD)

A Honeywell Company

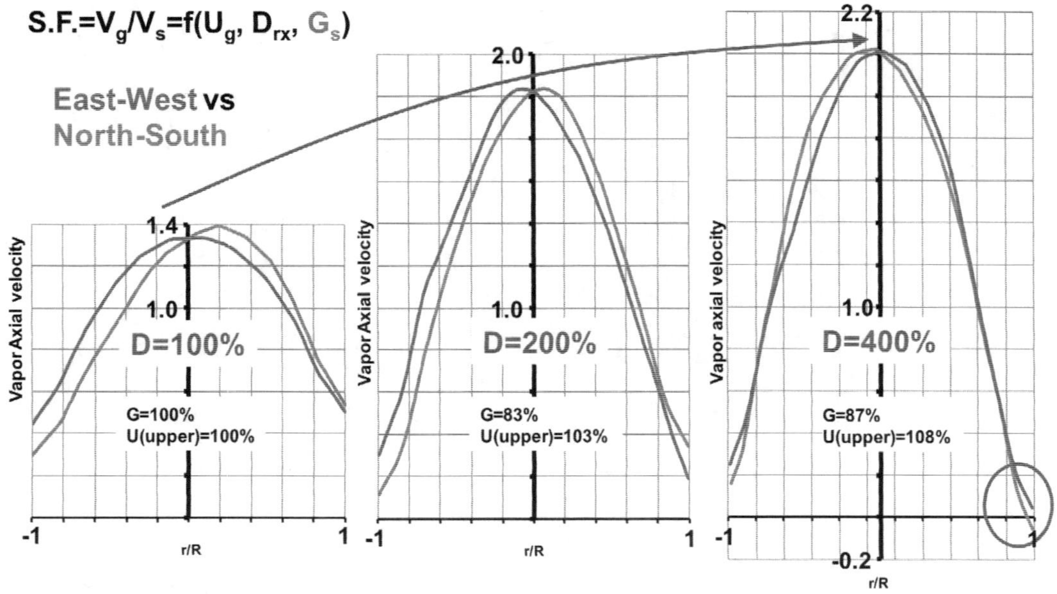

Larger riser diameter leads to triangular profile, even downflow

UOP 5978-14

Summary

uop
A Honeywell Company

- **Based on the CFD results the DPDPM results are closer to literature reported profiles**
- **Both models show the change in the flow profiles due to the change in gas velocity and catalyst flux**
  - As expected there is downflow at the wall with low gas velocity predicted by both models. This type of backmixing is not a preferred behavior in risers.
  - DPDPM predicts an inverted parabola for the flux profile with high gas velocity while the TFM does not
  - No downflow is observed in the high gas velocity case
  - DPDPM predicts reasonable density profiles for the high velocity case
- **DPDPM model requires a large number of simulated particles to produce reasonable results which sometimes is not feasible due to hardware limitations**

Particles Size using DPDPM model

UOP 5978-15

uop
A Honeywell Company

Q&A

UOP 5978-16

JOULE®

Photobioreactor Scale-Up

Michael D. Corbett
Process Engineering Manager
mcorbett@jouleunlimited.com

June 2013

©2013 Joule. Rights Reserved.

www.jouleunlimited.com

Overcoming the Obstacles to Truly Sustainable Fuels

JOULE®

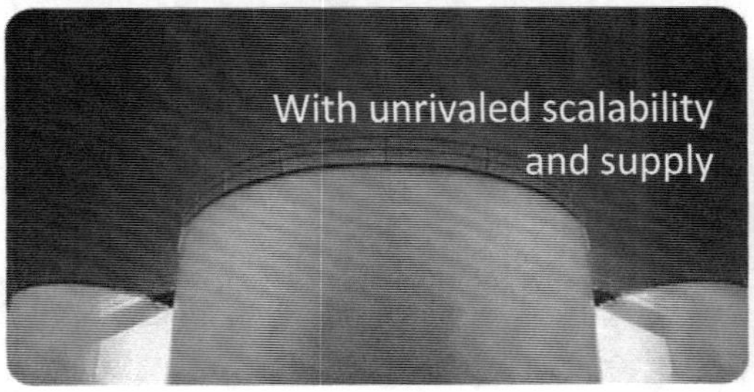

©2013 Joule. Rights Reserved.

2

www.jouleunlimited.com

A Direct and Continuous Conversion

Unlike processes using algae or other biomass feedstocks

JOULE®

FUEL	ENERGY SOURCE	CARBON SOURCE	FEEDSTOCK	PRE-TREATMENT PRODUCT	DOWNSTREAM PROCESS	PRODUCT	PRODUCTIVITY (gallons/acre/year)
BIOFUEL		BIOMASS	CORN	GLUCOSE	FERMENTATION	ETHANOL BUTANOL ALCOHOLS TERPENES	400 to 7,000 max
			CANE	SUCROSE			
			PLANT WASTE GRASS	GLUCOSE XYLOSE ARABINOSE MANNOSE, ETC			
			OIL PLANT CANOLA, ETC	TRIGLYCERIDE	EXTRACTION ESTERIFICATION	BIODIESEL ESTERS	
			ALGAE				
SOLAR FUELS		CO_2	DIRECT, CONTINUOUS SYNTHESIS			DIESEL ETHANOL CHEMICALS	**Ethanol Up to 25,000** **Diesel Up to 15,000**

©2013 Joule. Rights Reserved.

3

www.jouleunlimited.com

Abundant Resources = Localized Production Around the Globe

JOULE®

Worldwide Siting Opportunities

Sunlight

Ideal > $5 kWH/m^2/day$

CO_2

Power, cement, refinery, iron & steel companies

Land

- Non-arable land; near CO_2 source
- Relatively flat

Water

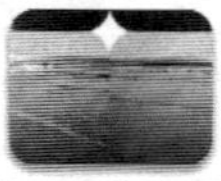

Waste, brackish, or sea water

Indicators of Scalability

Global Mean Solar Irradiance

> 2 billion tons CO_2 available globally in ideal sun locations[1]
> 2,000 sites with > 200,000 tons CO_2/year with ideal sunlight

700 Joule plants would consume 140 million tons of CO_2 equaling:
- < 0.5% of global CO_2 emissions
- 8% of global CO_2 emissions from cement companies

700 Joule plants equate to:
- < 1% of New Mexico or
- < 0.2% of global desert

- 97% of world's water is sea water
- Joule focusing on coastal locations
- Target industries often produce large amounts of waste water

[1] Source: IEA GHG CO2 Emissions Database v.2006 and NASA Solar database

©2013 Joule. Rights Reserved.

4

www.jouleunlimited.com

An Integrated Platform to Maximize Productivity

JOULE

The Microorganism

The *SolarConverter*® System

The *SunSprings*™ Plant

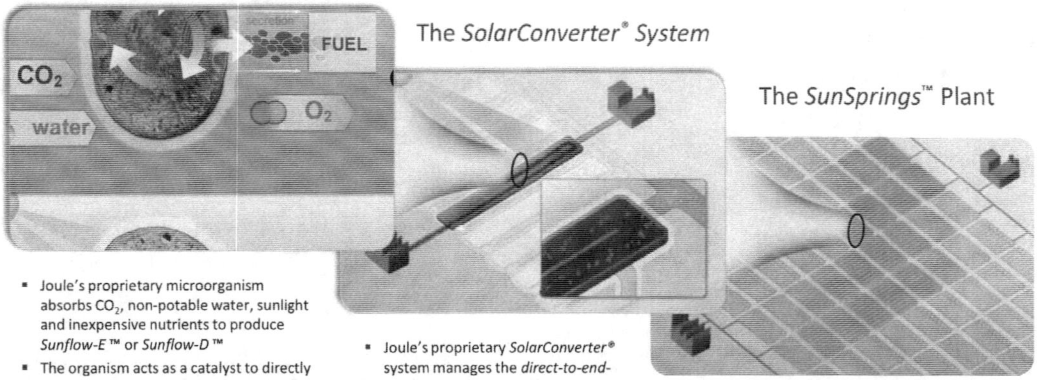

- Joule's proprietary microorganism absorbs CO_2, non-potable water, sunlight and inexpensive nutrients to produce *Sunflow-E* ™ or *Sunflow-D* ™
- The organism acts as a catalyst to directly and continuously *produce fuel*, not as an intermediate to produce lipids or sugars that are *subsequently converted to fuel.*
- Joule's inputs are readily available and truly sustainable, unlike processes dependent on agricultural or algal feedstocks

- Joule's proprietary *SolarConverter*® system manages the *direct-to-end-product-process,* including photon capture, CO_2 delivery and mixing, product synthesis and initial separation
- Production can be tailored to desired *Sunflow* ™ products, output levels, and various site sizes based on solar insolation, land and waste CO_2 availability

- Joule's *SunSprings*™ plants comprise modular circulation units that allow for cost-effective scale up
- Fuel production can start immediately upon the first unit deployment
- Plants have the flexibility to scale anywhere from 250 to 25,000 acres with similar unit level economics

©2013 Joule. Rights Reserved.

5

www.jouleunlimited.com

Cyanobacterium as a Production System

Combines optimum qualities for continuous production organism

JOULE

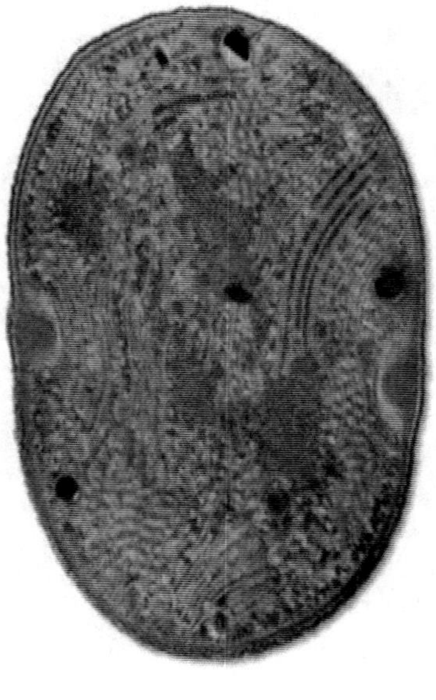

- Unicellular, Free-living
- No nuclear compartment
- Rapid metabolism
- Euryhaline: grows in brackish, saline, sea or fresh water
- Uses inexpensive medium components
- Takes up and integrates foreign DNA
- Engineerable thermotolerant species
- Functions for weeks in closed culturing systems
- No biofilm or fouling

©2013 Joule. Rights Reserved.

6

www.jouleunlimited.com

Reactor Design & Scale-Up

SolarConverter® Evolution

JOULE®

- CO2 Mass Transfer
- Light Management (Visible, IR, UV)
- Process Control
  - Temperature
  - pH
  - Mixing
  - OD
- Low Cost

Opportunity:
- Increase area coverage
- Reduce piping and connections
- Eliminate support structure
- Reduce amount of material
- Improve mixing

©2013 Joule. Rights Reserved.

Reactor Design & Scale-Up

Film based Photobioreactor - The most cost effective system for large area coverage

JOULE®

- Performance
  - Operates with higher than atmospheric CO2 concentrations
  - Closed system to minimize contamination
  - Minimizes volume of water vs. deep pond or raceway approaches
- Logistics Supply Chain
  - Reduces volume of polymer resin used
  - Minimizes shipping cost - shipping collapsible film vs. rigid parts
  - Leverages existing high volume manufacturing methods
- Field Deployment at Scale
  - Collapsible capsules can be unrolled easily for deployment
  - Flexible, non-rigid capsules are less susceptible to damage during deployment
  - Replaceable
  - Can fit into existing recycling streams

©2013 Joule. Rights Reserved.

Reactor Scale-Up

Lab Scale to Commercial Scale

JOULE®

©2013 Joule. Rights Reserved.

www.jouleunlimited.com

Reactor Design & Scale-Up

Utilizing experimental and calculation based modeling techniques to optimize design

JOULE®

Computational Fluid Dynamics

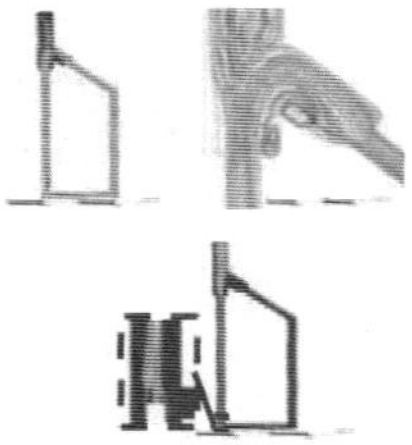

Chemical Process Simulator

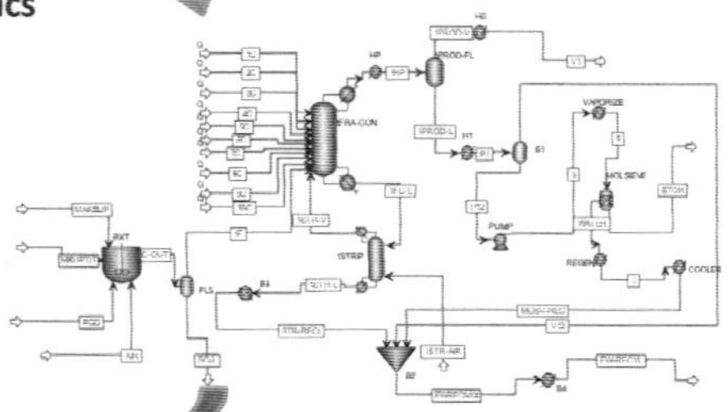

©2013 Joule. Rights Reserved.

www.jouleunlimited.com

Reactor Design & Scale-Up

Using CFD modeling in as a cost effective alternative to experimental modeling

JOULE

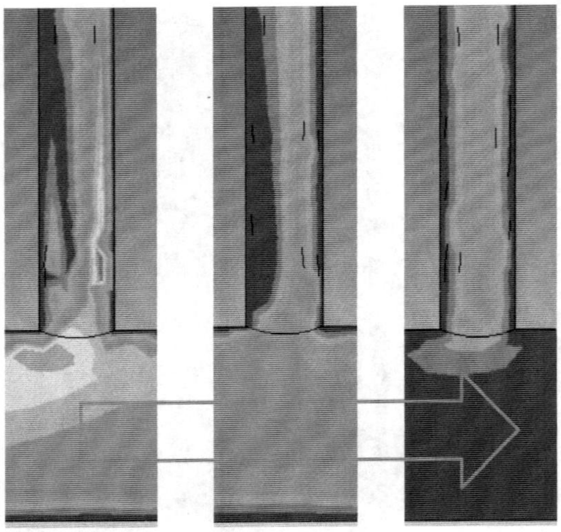

Velocity Contours

Branch 1 Branch 7 Branch 14

Flow Distribution

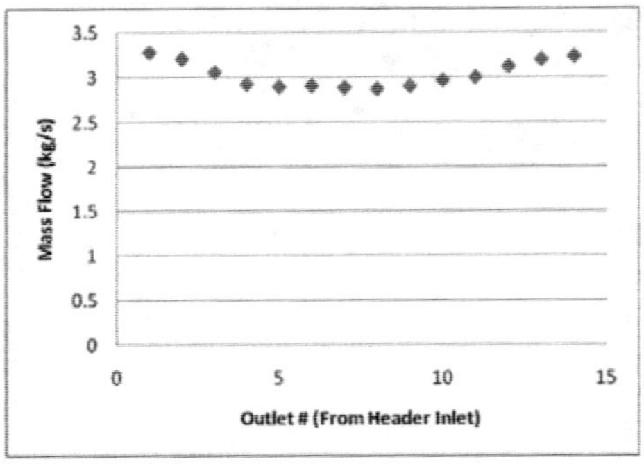

©2013 Joule. Rights Reserved.

Reactor Design & Scale-Up

Matlab as a tool for parametric design studies

JOULE

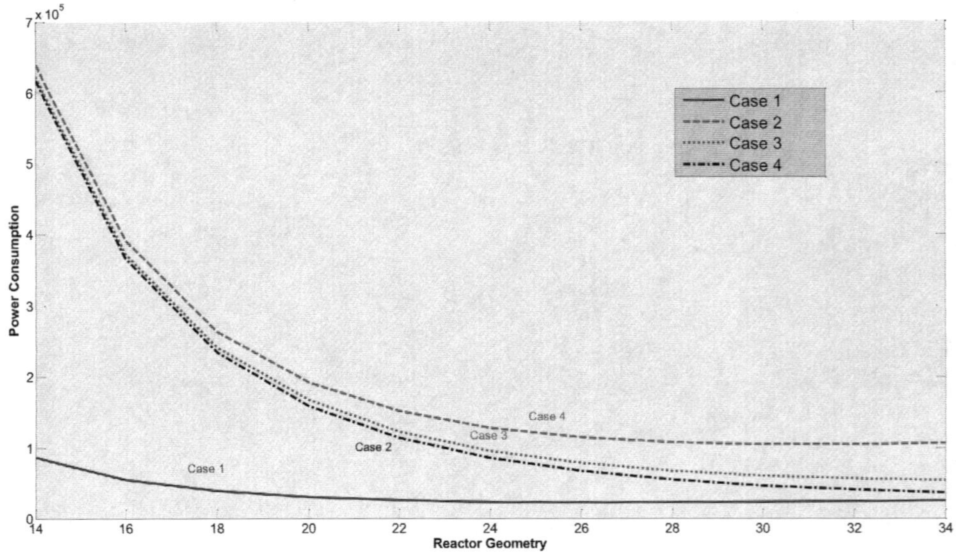

©2013 Joule. Rights Reserved.

Reactor Scale-Up

PBR design that cycles cells through photic and dark zones matches energy conversion rates JOULE

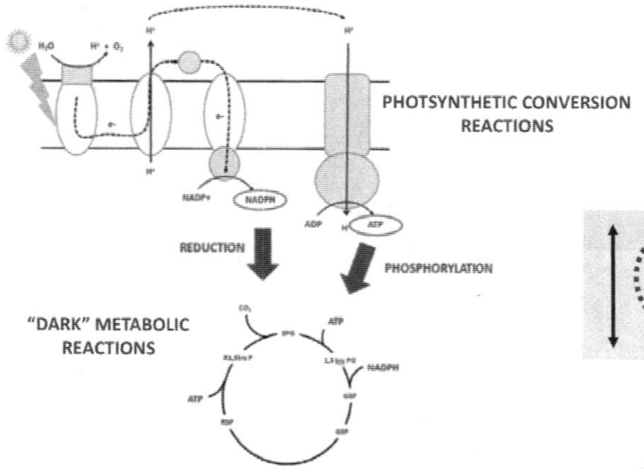

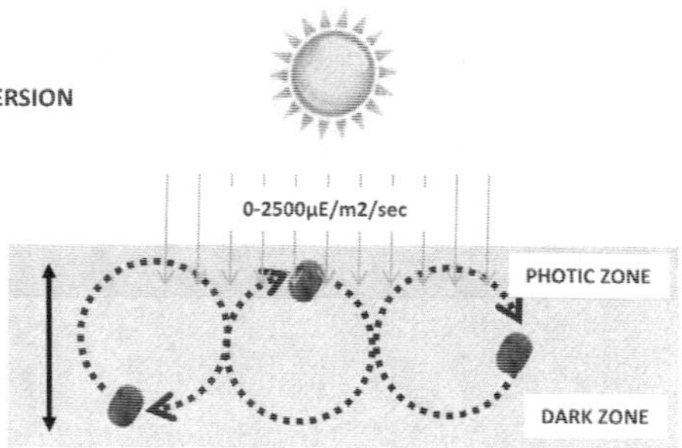

- Light reactions convert photons to chemical energy
- Dark reactions use chemical energy to drive metabolism
- Mismatch leads to over-reduction and photoinhibition

- Cycling cells through insolation and metabolic periods prevents over-reduction and photoinhibition
- Cycling can be accomplished via mixing of high OD culture through photic/non-photic zones

©2013 Joule. Rights Reserved.

www.jouleunlimited.com

Reactor Scale-Up

Current SolarConverter® System in Operation JOULE

SolarConverter Features

- A simple, low cost, manufacturable design for efficient area coverage
- High performance materials with exceptional outdoor performance
- Precise process control for optimum productivity
  - Temperature control
  - Flow rate for optimum mixing
  - Light management
- Cleanable, modular process design that easily scales to thousands of acres
- High media flow rates for superior mixing already achieved
- A high performance reactor vessel constructed using outdoor-tough, low-cost polymeric films

©2013 Joule. Rights Reserved.

14

www.jouleunlimited.com

Reactor Scale-Up
Simultaneous development & reactor design

JOULE®

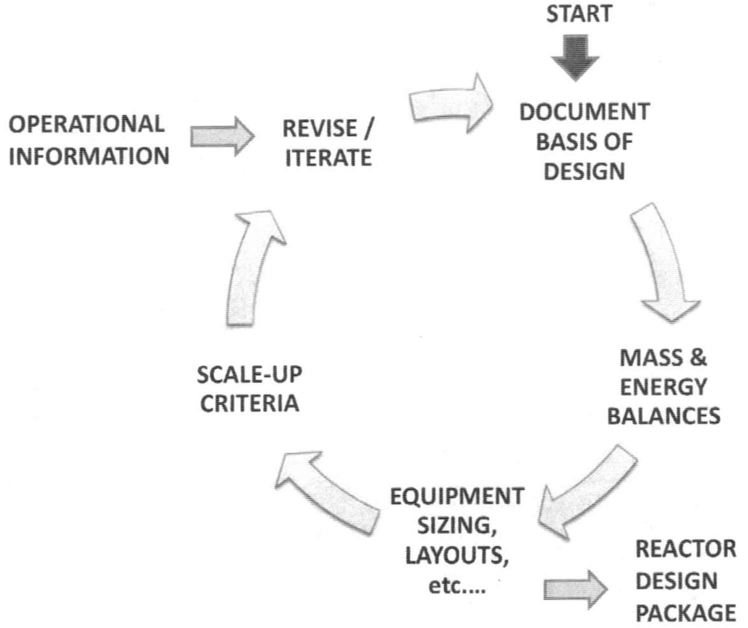

©2013 Joule. Rights Reserved. 15 www.jouleunlimited.com

Reactor Scale-Up
SolarConverter® 10X Scale-up in Construction Phase

JOULE®

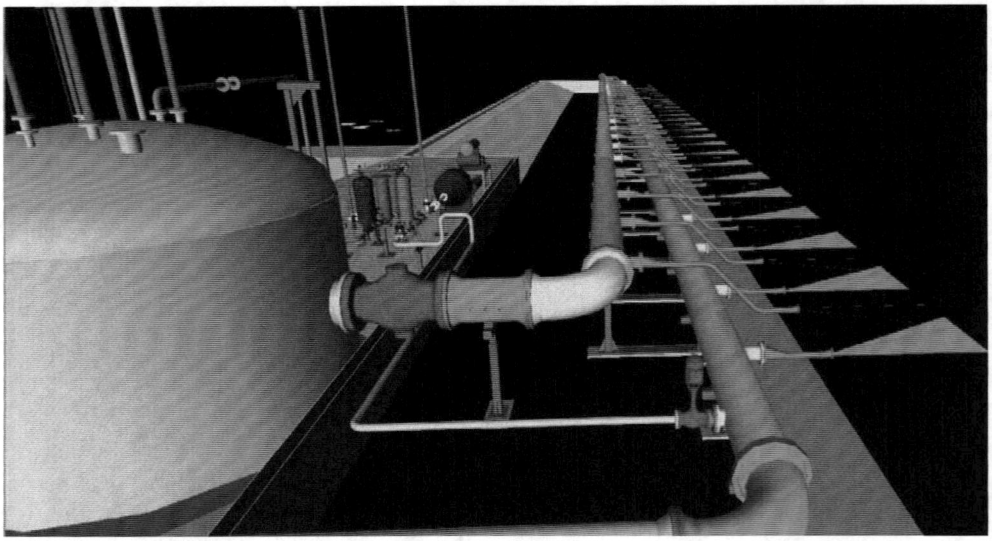

©2013 Joule. Rights Reserved. 16 www.jouleunlimited.com

Commercial Facility Scale-Up

JOULE®

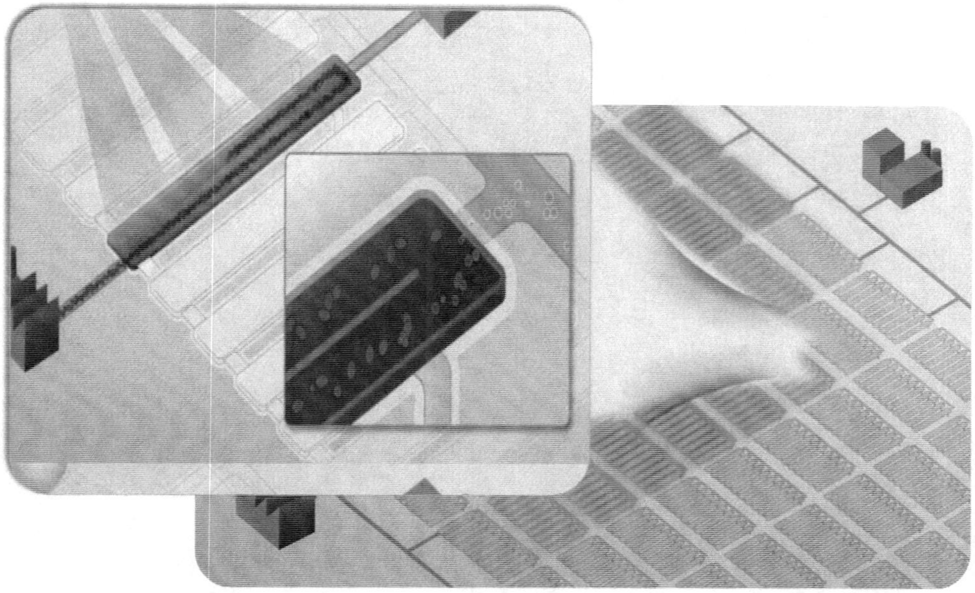

©2013 Joule. Rights Reserved. 17 www.jouleunlimited.com

Commercial Facility Scale-Up

Value of getting it "close" now

JOULE®

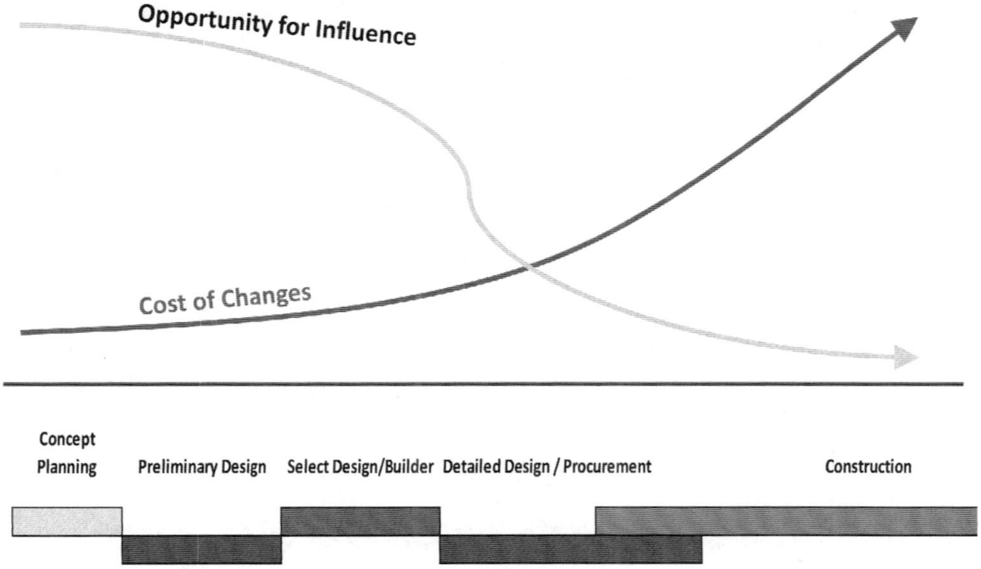

©2013 Joule. Rights Reserved. 18 www.jouleunlimited.com

Commercial Facility Scale-Up

Process modeling - need for a creative solution

JOULE®

- 300,000 bpd Oil Refinery
  - Complex simulation required
  - Feed is relatively constant throughout the year

vs.

- 1,000 Acre Joule Plant
  - Complex simulation required
  - **Feed varies every hour of the year!**

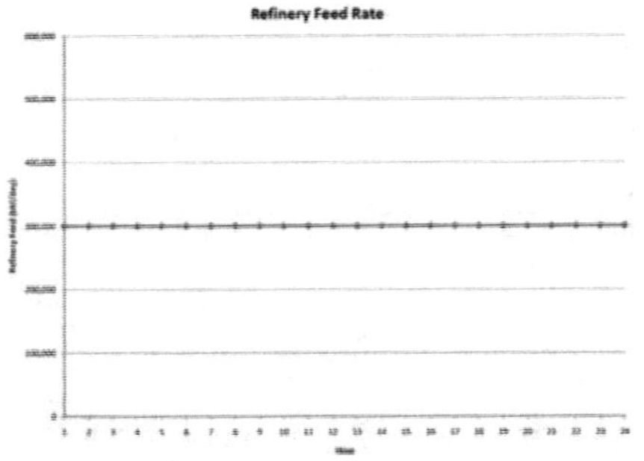

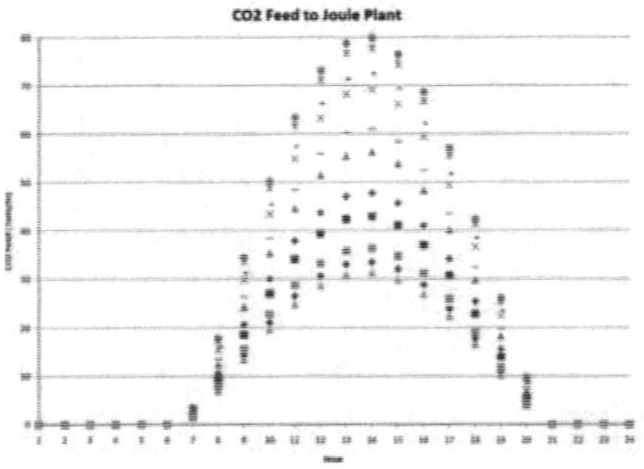

Cannot easily use commercially available simulators! → **Developed proprietary in-house software platform.**

©2013 Joule. Rights Reserved. 19 www.jouleunlimited.com

Commercial Facility Scale-Up

Integrated facility optimization via in-house developed model

JOULE®

- Individual Optimizations
  - SolarCoverter® Parameters
  - Biological Parameters
- System Optimizations
  - Thermal Management
  - O_2 / Product Stripping
- Sensitivity Analysis
  - Site Specific Temperature Data
  - Insolation Data

Cost Effective Commercial Scale Facility Design

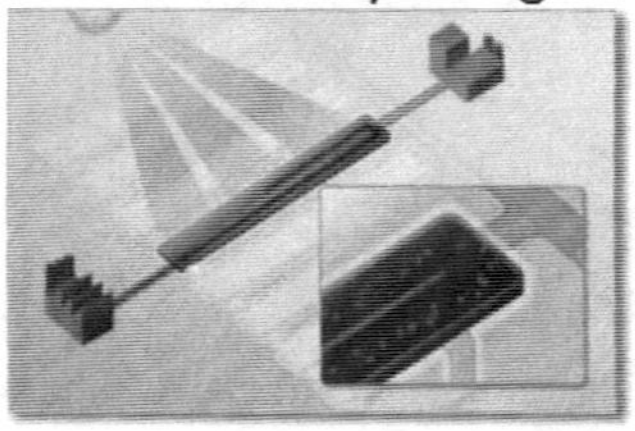

©2013 Joule. Rights Reserved. 20 www.jouleunlimited.com

Questions?

SunSprings™ Hobbs: demonstrating commercial readiness

JOULE

Scalability of CHER Reactor from Bench to Commercial Scale

Jia, Z., Chart Energy & Chemicals

Originally presented on: 6/13/2013 14:00:00 - 14:30:00

Chart Energy and Chemicals has developed the Compact Heat Exchange Reactor (CHER) based on its brazed aluminum heat exchanger technology. CHER has excellent heat transfer properties, allowing for process intensification through increased reaction rate, process integration of reaction and heat transfer, and a more compact reactor design. Because of the repeating layered structure and flexible reactor design, CHER can be easily scaled up to the desired capacity. Scalability of CHER is performed by choosing the proper reactor size and capacity as well as by modularized reactor units, depending on the maximum target plant productivity and desired operational flexibility such as a wider range of operation.

Chart Energy and Chemicals has developed CHER for small- and mid-scale Fischer-Tropsch application. The Fischer-Tropsch reaction is highly exothermic and proper temperature control is essential for sustained operation of XTL plants. This presentation will show Chart's development from bench-scale reactors to current pilot scale demonstration. Reactor performance data for both scales will be presented, and commercial scale design will also be discussed.

Do's and Don'ts of Fluidized Bed Scale-Up

Ray Cocco, Ted Knowlton and S.B. Reddy Karri
Particulate Solid Research, Inc. (PSRI)

June 13, 2013

PSRI Particulate Solid Research, Inc. *Applying the Fundamentals*

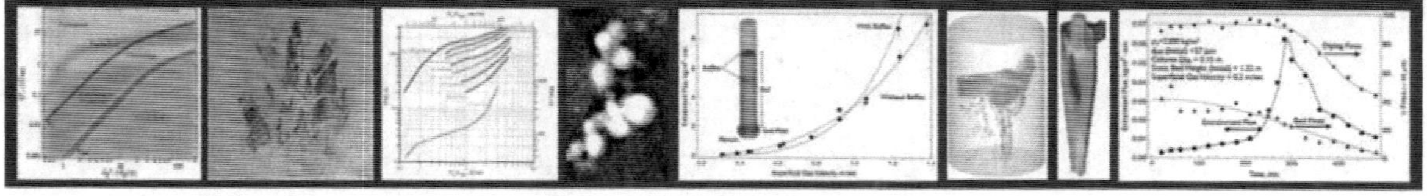

Granular-Fluid Flow Behavior

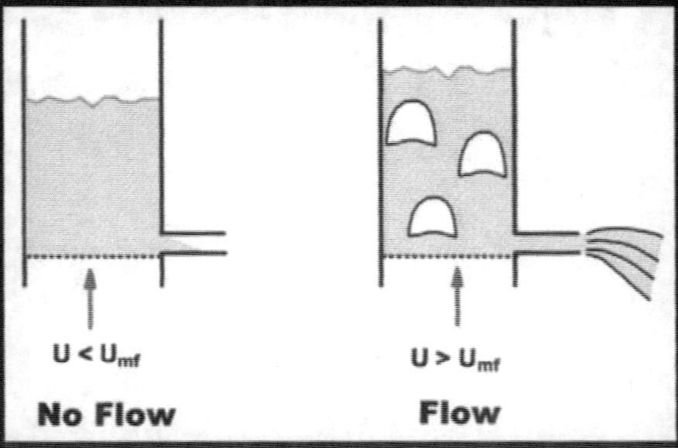

- Fluidization is the operation by which solid particles are transformed into a fluidlike state through suspension in a gas or liquid*.

*FROM D. KUNII, O. LEVENSPIEL, FLUIDIZATION ENGINEERING, 2ND, BUTTERWORTH-HEINEMANN, 1991

PSRI *Applying the Fundamentals*

Achieving Fluidization

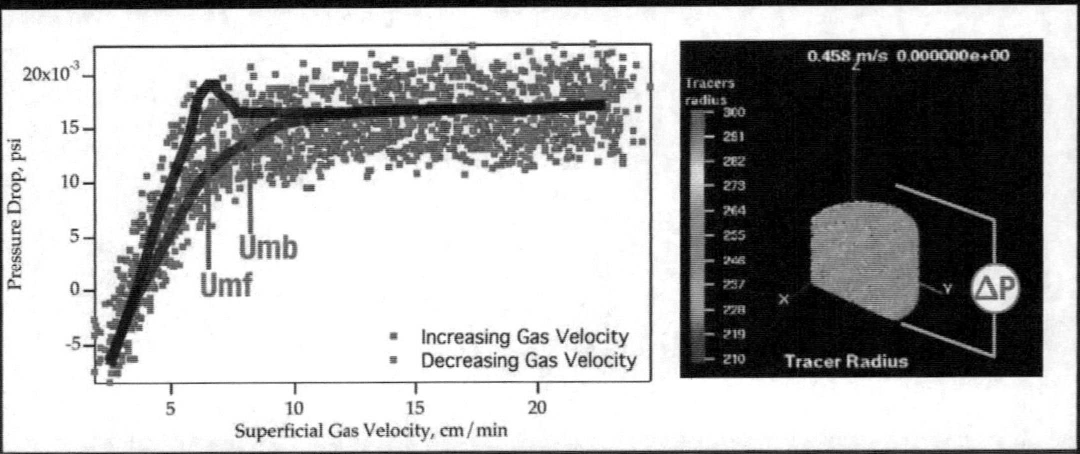

- Fluidization occurs when the drag force of the gas on the particles exceeds the weight of the particles

Fluidization Regimes

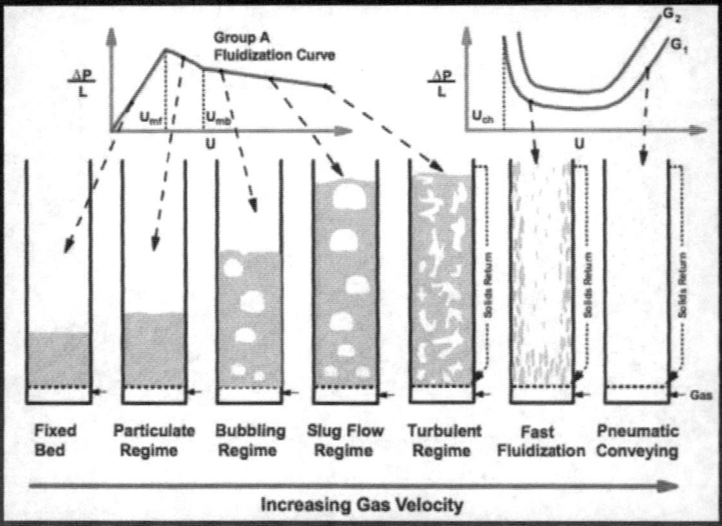

Particle Properties and Fluidization

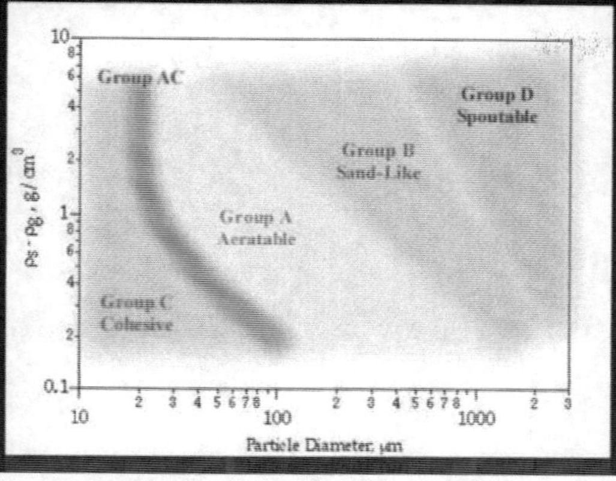

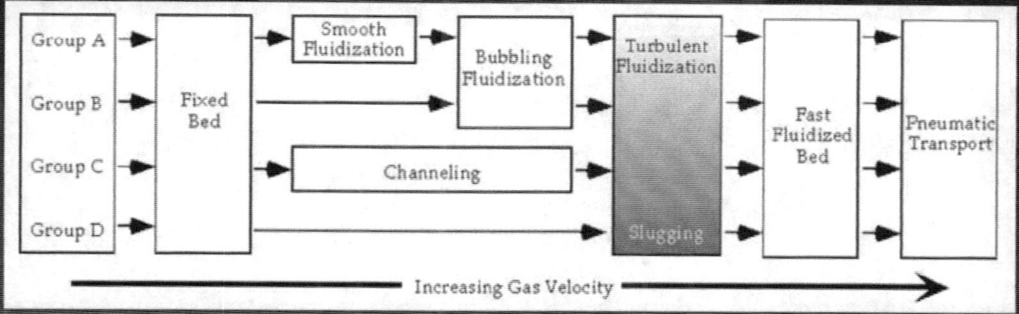

Components of a Fluidized Bed

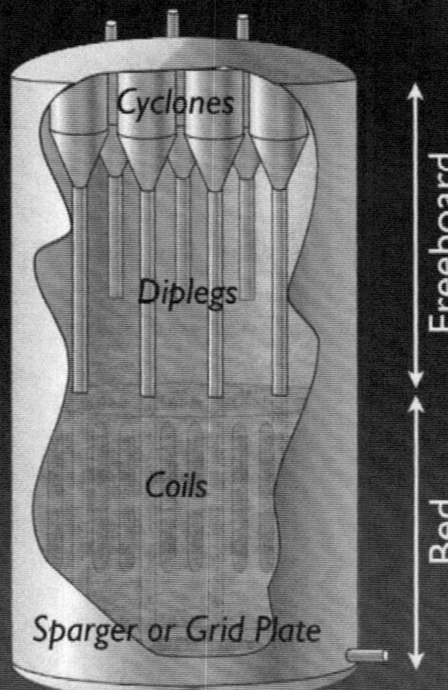

- Sparger or grid plate (gas distributor) for uniform distribution of feed gas
- Coils for heating or cooling
- Cyclones for gas-solid separation
- Diplegs for solids return to the fluidized bed

Introduction > Fluidized Bed Applications

Fluidized Bed Applications

- 1920's
  - Winkler coal gasification
- 1940's
  - Fluid catalytic cracking (Houdry FCC)
  - Phthalic anhydride
  - Ore Roasting
  - Drying
- 1950's
  - Fluid hydroforming
  - Fischer-Tropsh
  - Acrylonitrile
  - Spouted bed
- 1960's
  - Three-phase biochemical processes
- 1970's
  - Circulating fluidized beds
  - FCC risers
- 1980's
  - Polyethylene and polypropylene
  - Coal combustors
- 1990's
  - Short contact time reactors
  - Improved coal gasification
- 2000's
  - Fine chemical applications
- 2010's
  - Polycrystalline silicons
  - Biomass

PSRI Applying the Fundamentals

Design Concerns

Fluidized Bed Design Concerns

- Bubble size
- Gas distribution
- Entrainment
- Cyclone reliability

PSRI Applying the Fundamentals

Why Particle Size is Important?

Geldart Group A Particles

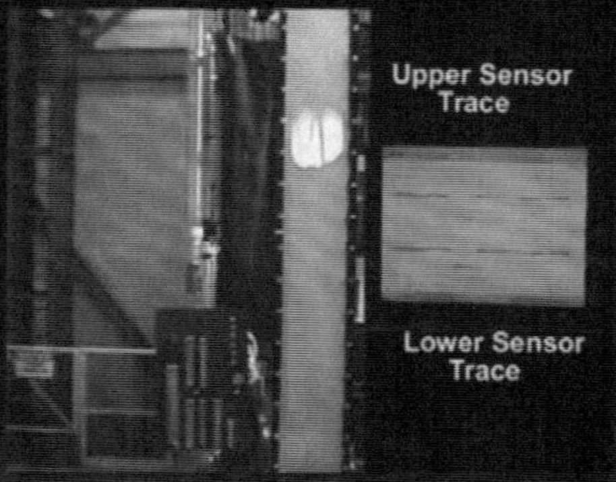

Geldart Group B Particles

- Smaller bubbles provide higher heat and mass transfer in the bed than larger bubbles
- Larger bubbles have lower surface-to-volume ratios
- Larger bubbles travel faster through a bed than smaller bubbles
- Pressure and particle size play an important role in bubble size
- Large bubbles more likely to slug even in commercial units
- Fines levels can be important too

Gas Distribution

- Basically two types of distributing devices
  - Spargers
  - Grid plates
- Design constraints
  - Orifice pressure drop should be> 1/3X bed pressure drop for tall beds
  - 1X bed pressure drop for shallow beds
  - Manifold velocities
  - Catalyst weeping
  - Attrition from jets

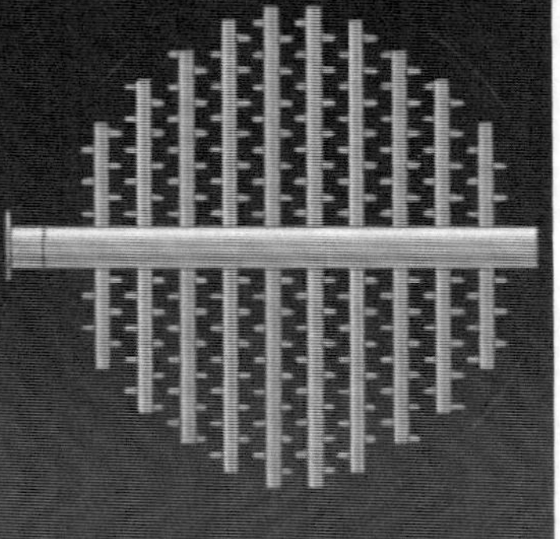

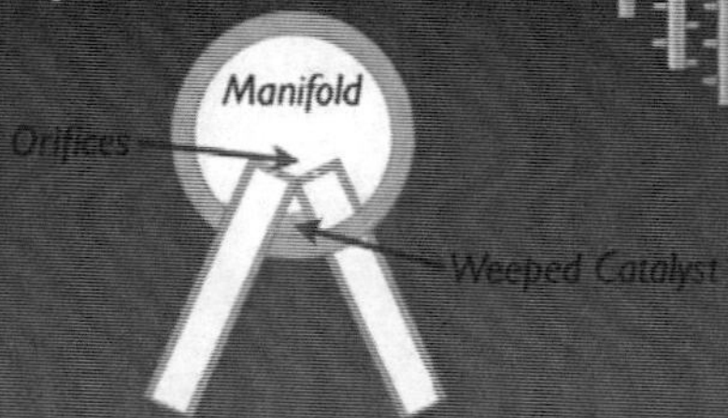

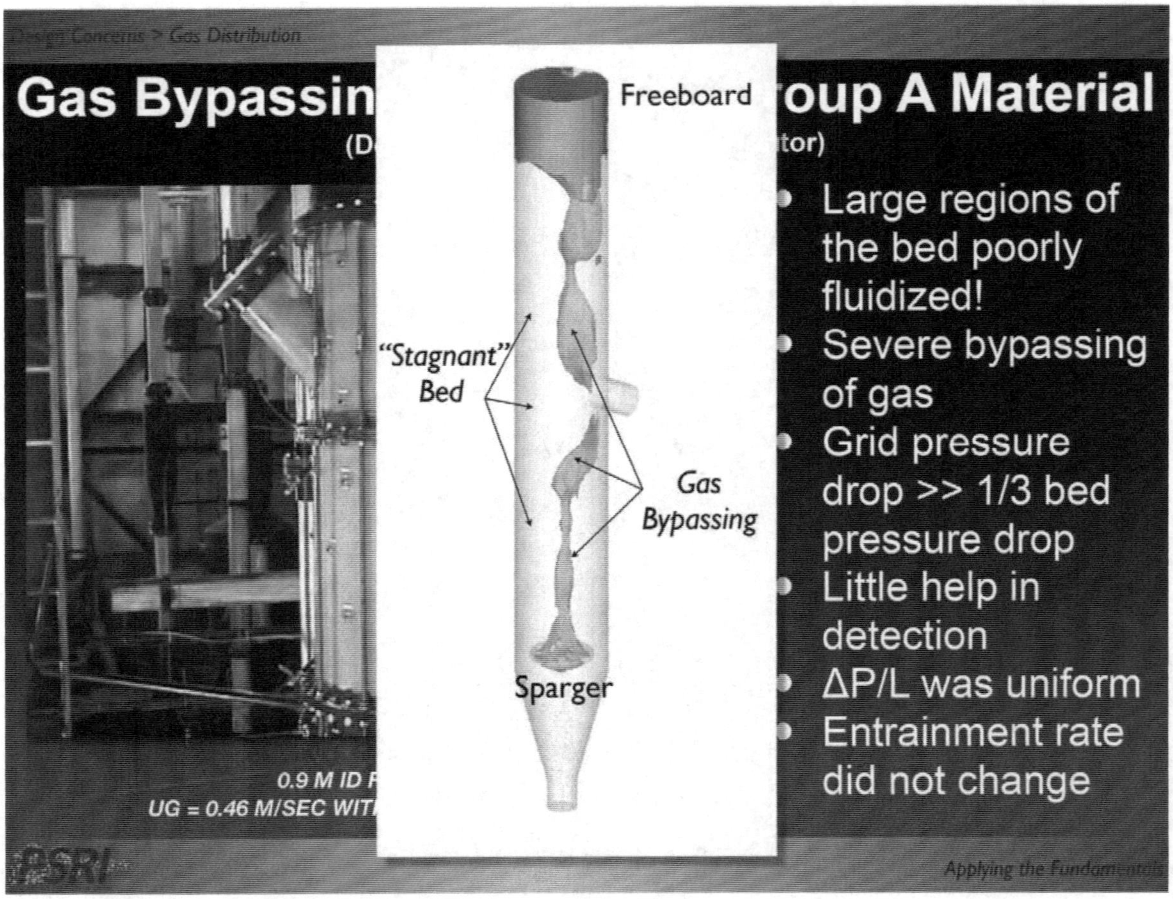
Design Concerns > Gas Distribution
Gas Bypassin
oup A Material
(D
tor)
Freeboard
"Stagnant" Bed
Gas Bypassing
Sparger
Large regions of the bed poorly fluidized!
Severe bypassing of gas
Grid pressure drop >> 1/3 bed pressure drop
Little help in detection
ΔP/L was uniform
Entrainment rate did not change
0.9 M ID F
UG = 0.46 M/SEC WIT
PSRI
Applying the Fundamentals

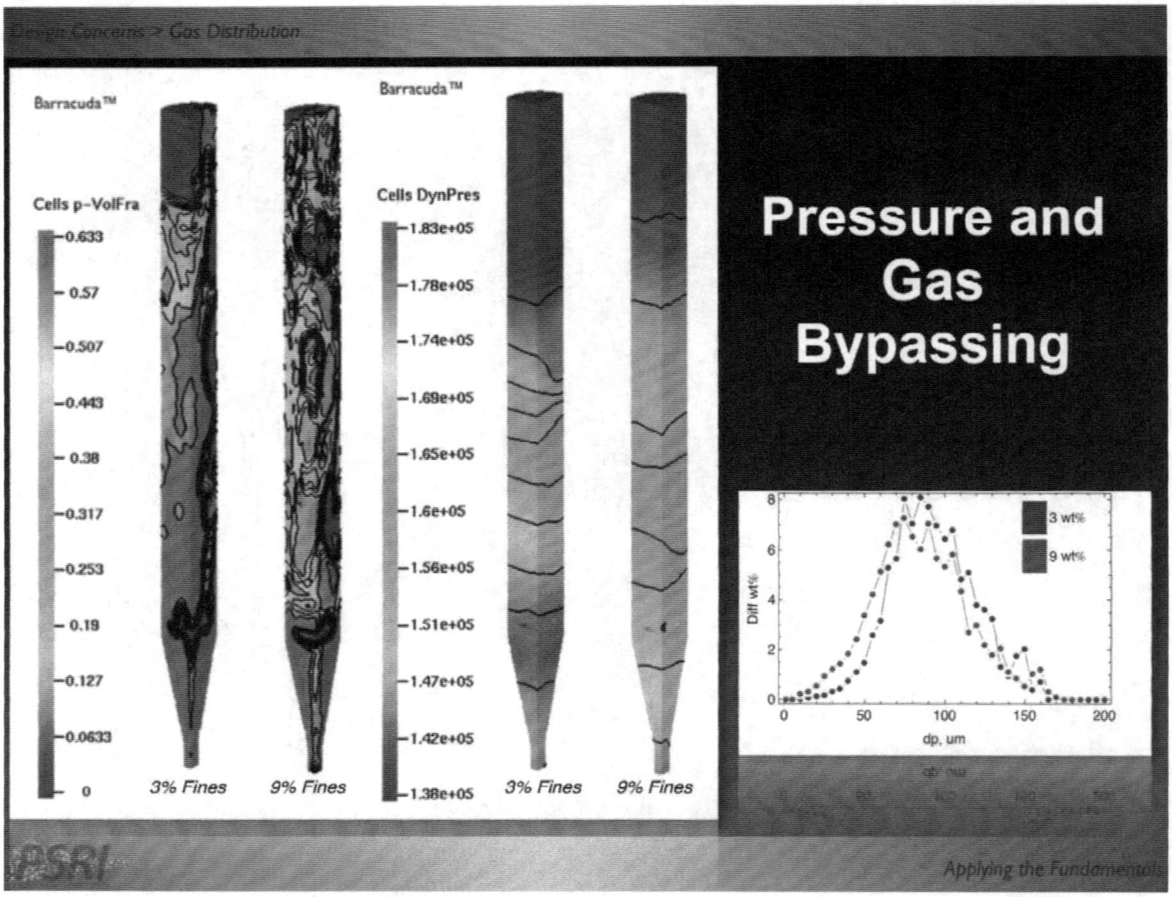
Design Concerns > Gas Distribution
Barracuda™
Cells p-VolFra
0.633
0.57
0.507
0.443
0.38
0.317
0.253
0.19
0.127
0.0633
0
3% Fines
9% Fines
Barracuda™
Cells DynPres
1.83e+05
1.78e+05
1.74e+05
1.69e+05
1.65e+05
1.6e+05
1.56e+05
1.51e+05
1.47e+05
1.42e+05
1.38e+05
3% Fines
9% Fines
Pressure and Gas Bypassing
Diff wt%
dp, um
3 wt%
9 wt%
PSRI
Applying the Fundamentals

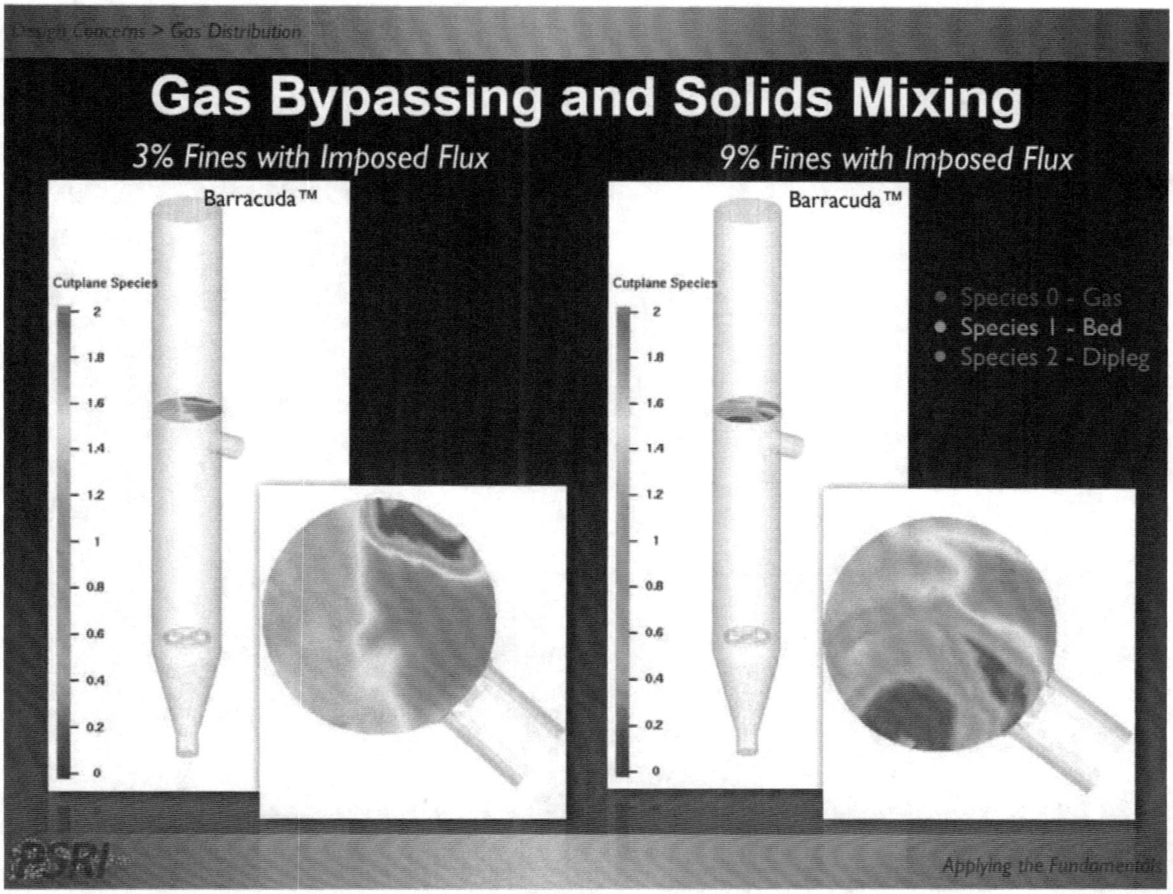

Design Concerns > Gas Distribution

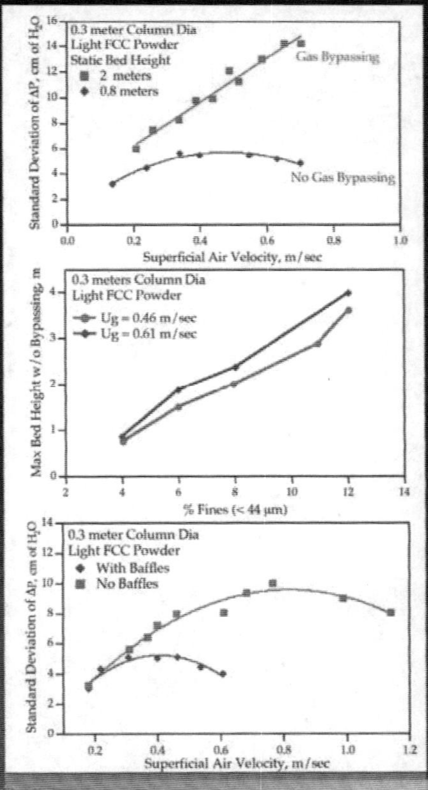

Managing Gas Bypassing

- Jet streaming is a function of gas permeability and bed pressure drop (i.e., weight)
  - Mostly with Geldart Group A powders
- Jet streaming can be managed
  - Limiting the bed height
    - Not always possible
  - Adding particle fines
  - Increasing the pressure
    - More gas can get into the emulsion
- Adding baffles

PSRI

Applying the Fundamentals

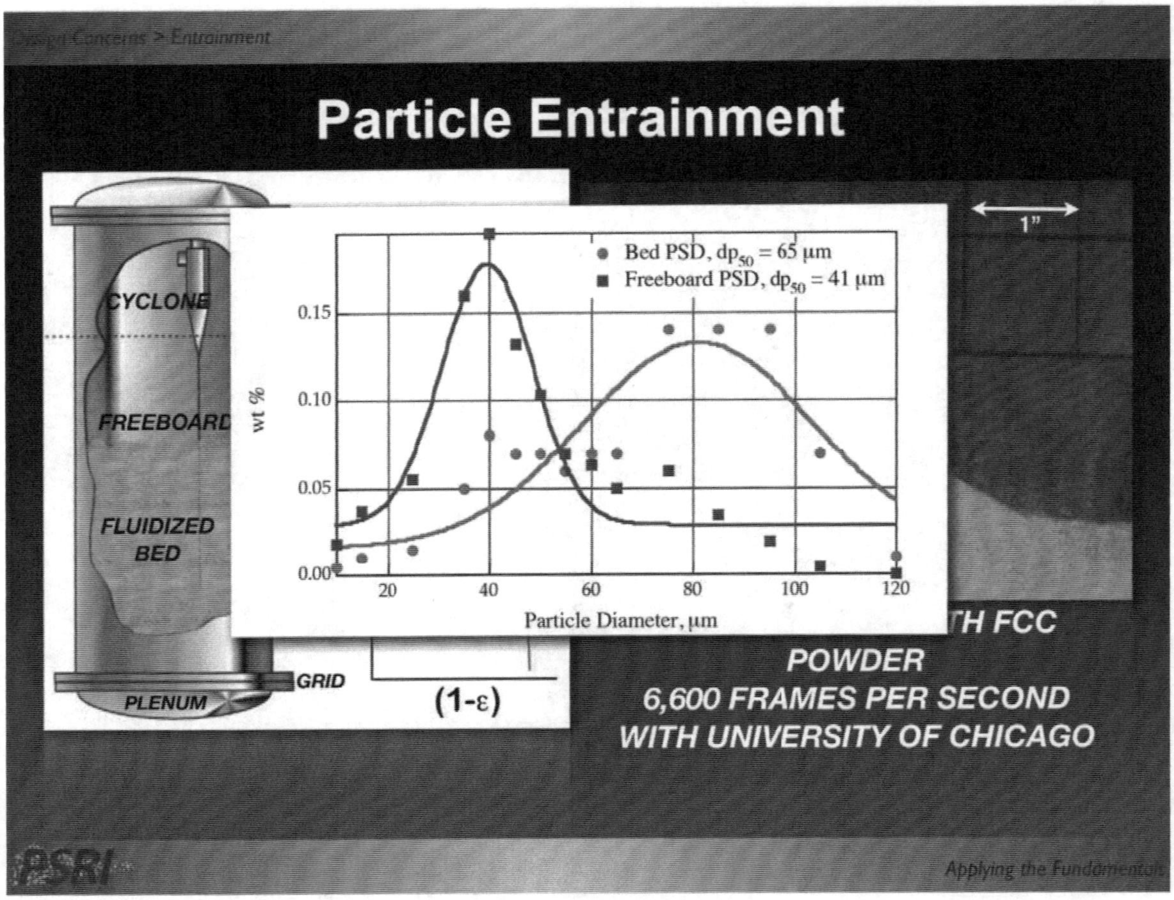
Design Concerns > Entrainment
Particle Entrainment
CYCLONE
FREEBOARD
FLUIDIZED BED
GRID
PLENUM
(1-ε)
1"
Bed PSD, dp_{50} = 65 μm
Freeboard PSD, dp_{50} = 41 μm
wt %
0.15
0.10
0.05
0.00
20
40
60
80
100
120
Particle Diameter, μm
TH FCC
POWDER
6,600 FRAMES PER SECOND
WITH UNIVERSITY OF CHICAGO
PSRI
Applying the Fundamentals

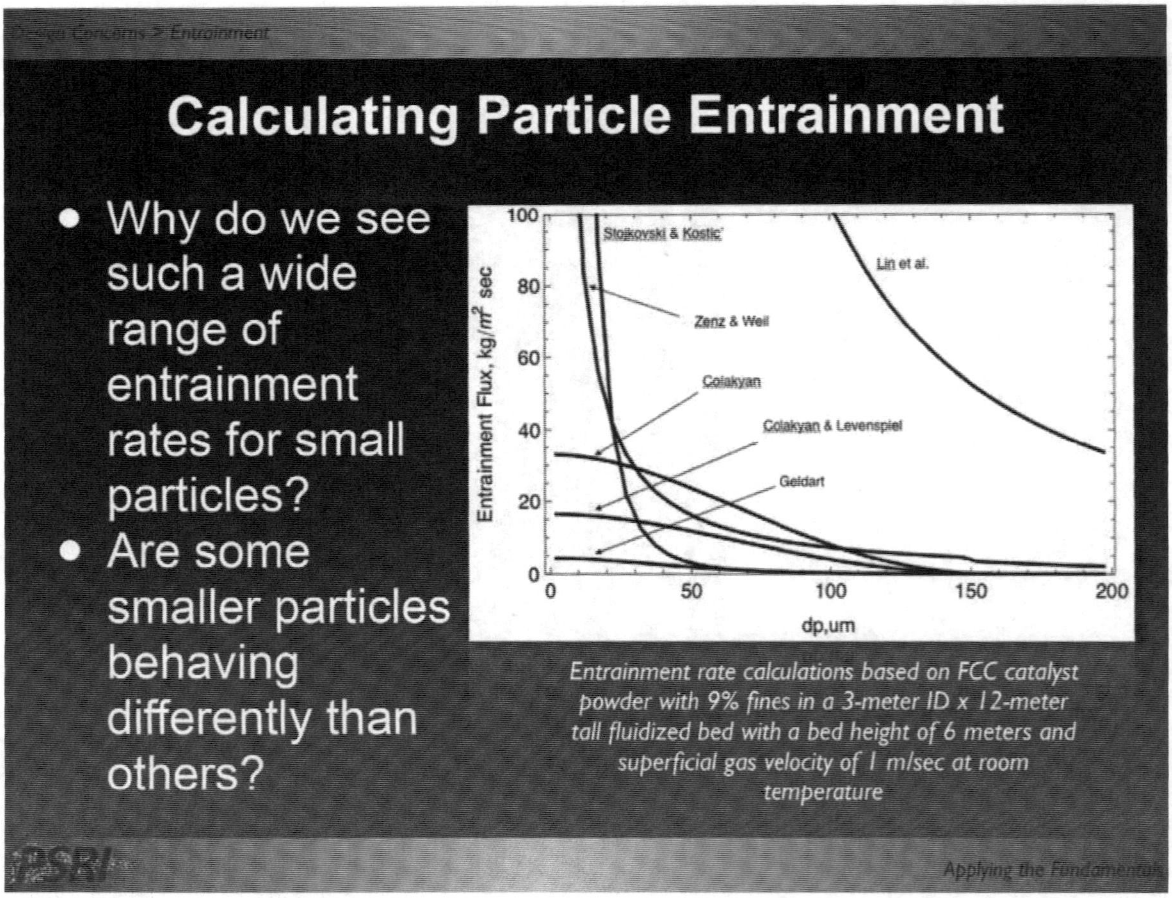
Design Concerns > Entrainment
Calculating Particle Entrainment
• Why do we see such a wide range of entrainment rates for small particles?
• Are some smaller particles behaving differently than others?
Stojkovski & Kostic'
Lin et al.
Zenz & Weil
Colakyan
Colakyan & Levenspiel
Geldart
Entrainment Flux, kg/m² sec
100
80
60
40
20
0
0
50
100
150
200
dp,um
Entrainment rate calculations based on FCC catalyst powder with 9% fines in a 3-meter ID x 12-meter tall fluidized bed with a bed height of 6 meters and superficial gas velocity of 1 m/sec at room temperature
PSRI
Applying the Fundamentals

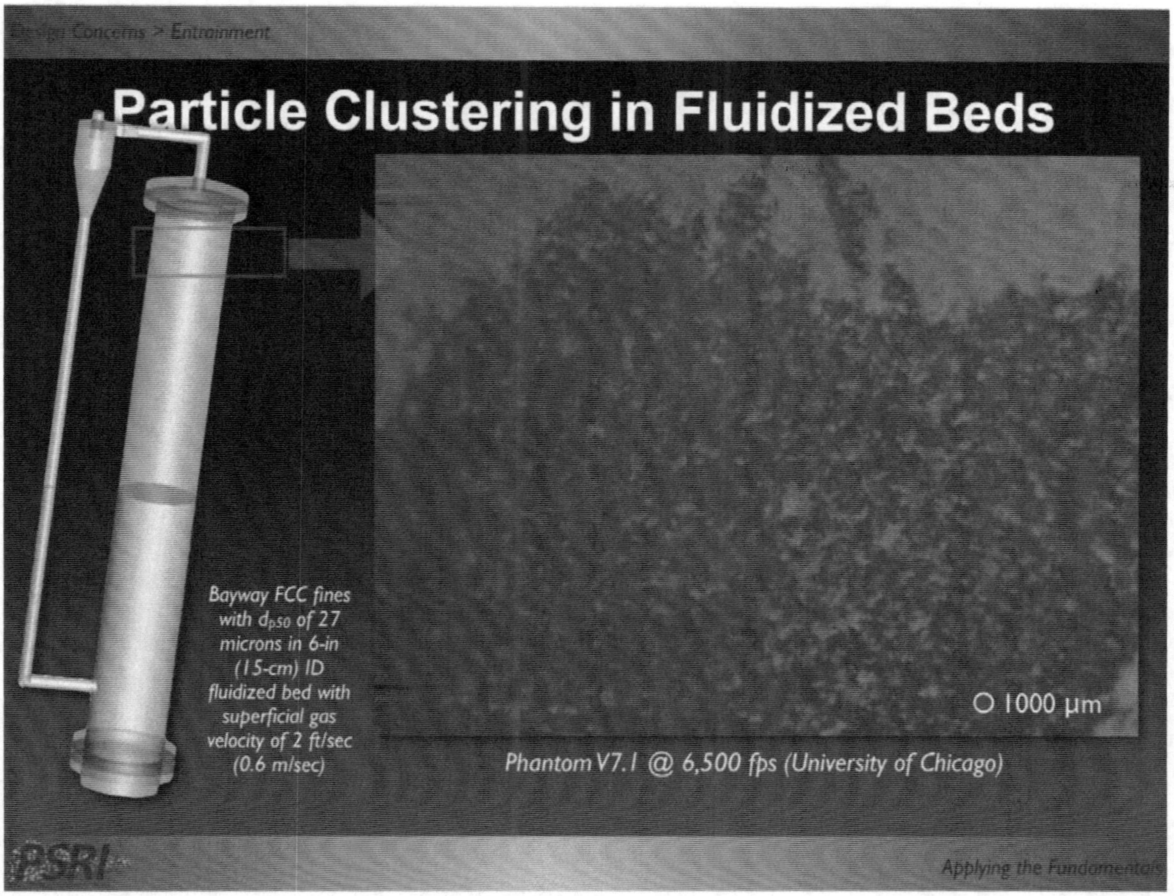
Concerns > Entrainment
Particle Clustering in Fluidized Beds
Bayway FCC fines with d_{p50} of 27 microns in 6-in (15-cm) ID fluidized bed with superficial gas velocity of 2 ft/sec (0.6 m/sec)
O 1000 µm
Phantom V7.1 @ 6,500 fps (University of Chicago)
PSRI
Applying the Fundamentals

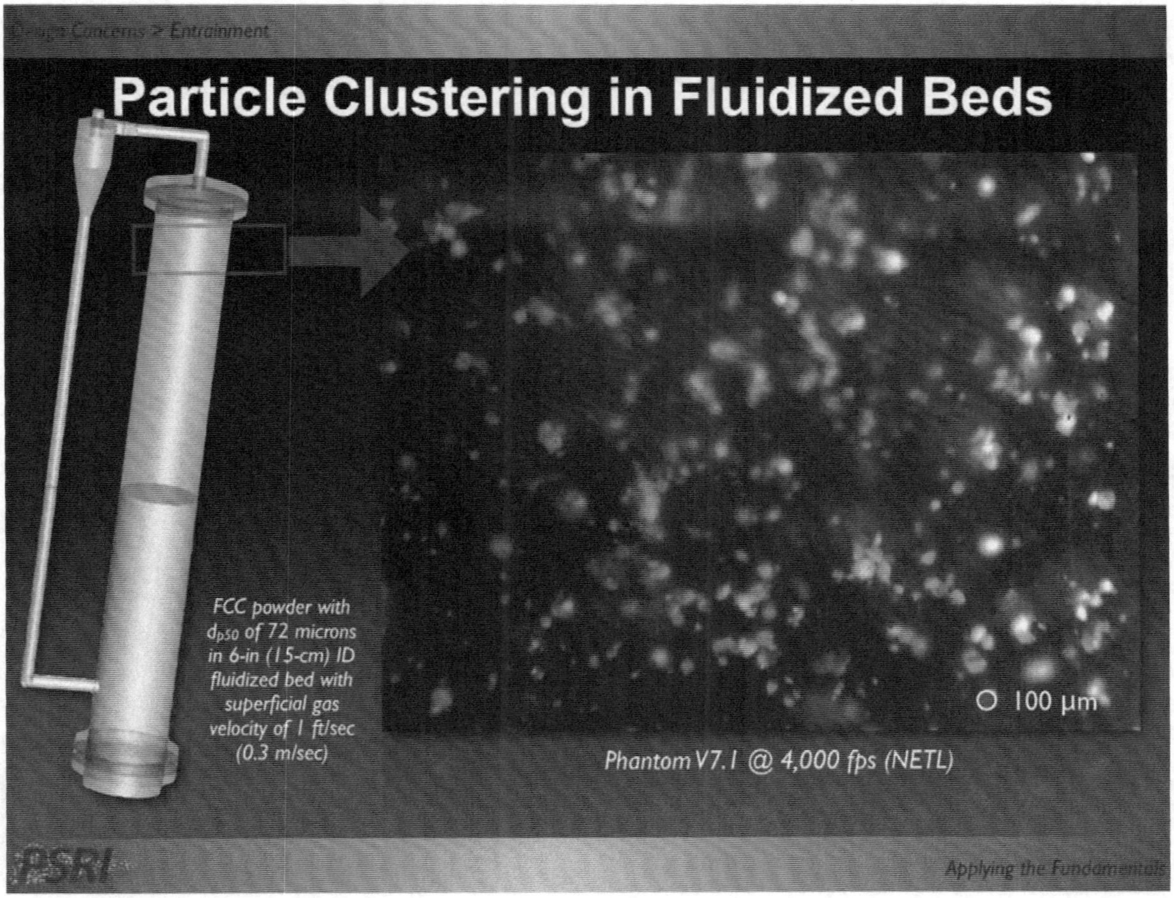
Concerns > Entrainment
Particle Clustering in Fluidized Beds
FCC powder with d_{p50} of 72 microns in 6-in (15-cm) ID fluidized bed with superficial gas velocity of 1 ft/sec (0.3 m/sec)
O 100 µm
Phantom V7.1 @ 4,000 fps (NETL)
PSRI
Applying the Fundamentals

Design Concerns > Entrainment
Particle Clustering in Fluidized Beds
FCC powder with d_{p50} of 72 microns in 6-in (15-cm) ID fluidized bed with superficial gas velocity of 1 ft/sec (0.3 m/sec)
100 µm
Phantom V7.1 @ 4,000 fps (NETL)
PSRI
Applying the Fundamentals

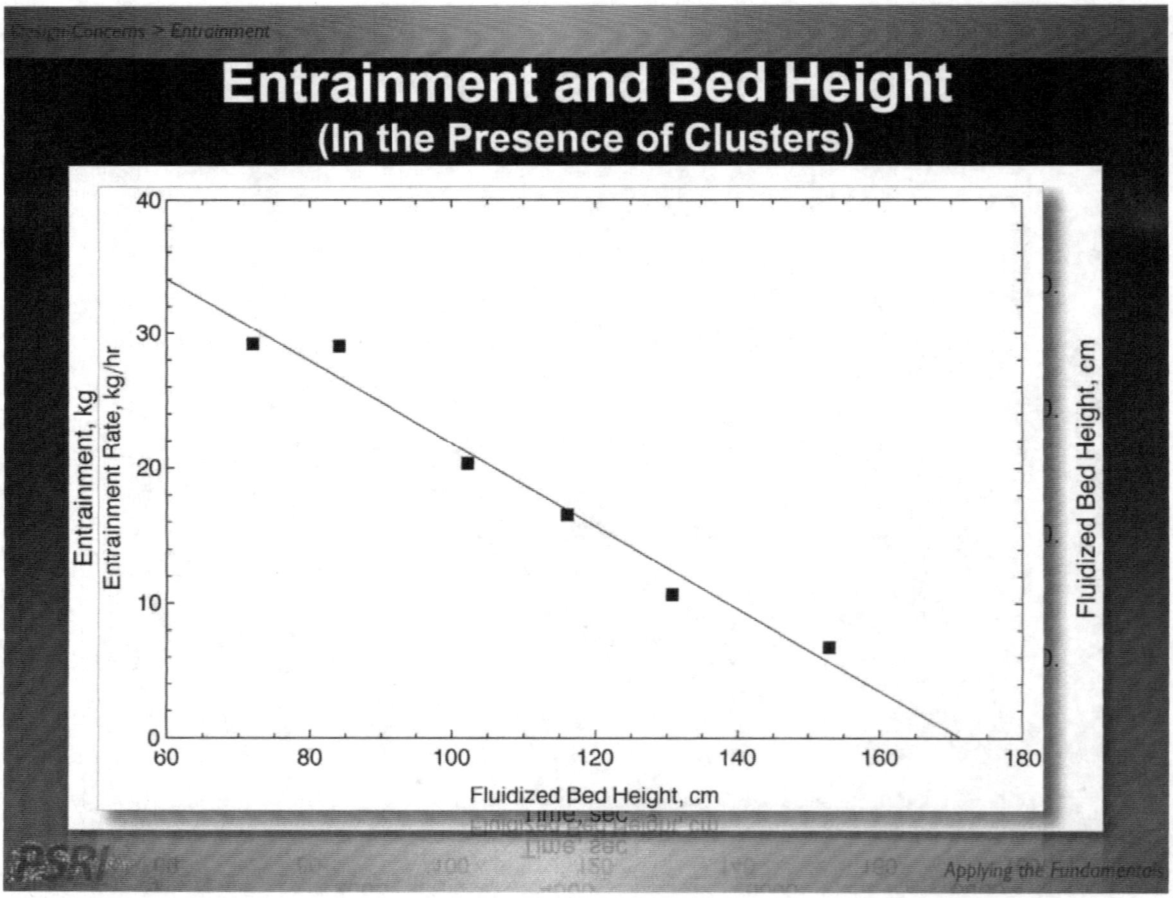

Design Concerns > Entrainment
Entrainment and Bed Height
(In the Presence of Clusters)
Entrainment, kg
Entrainment Rate, kg/hr
0
10
20
30
40
60
80
100
120
140
160
180
Fluidized Bed Height, cm
Fluidized Bed Height, cm
Time, sec
PSRI
Applying the Fundamentals

Baffles and Entrainment
(In the Presence of Clusters)

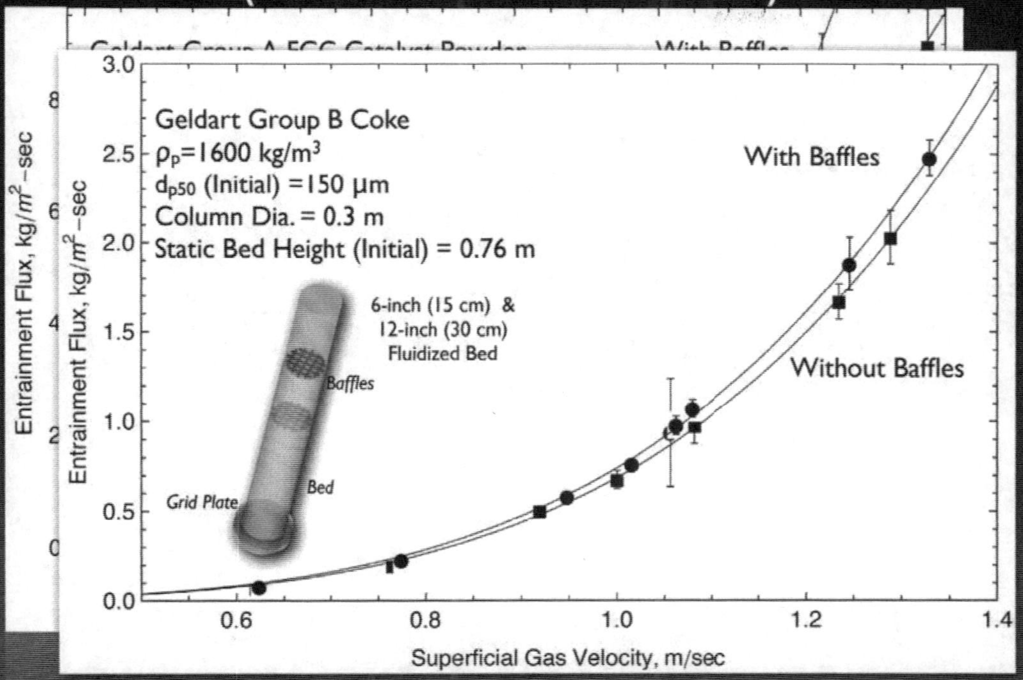

Cluster Formation and Stability in Fluidized Beds

- Particle clusters may form near the bottom of the bed and continue to grow as they migrate to the top of the bed, possibly with the help of bubbles
- At the top of the bed, clusters are either entrained or circulate back down to the bottom of the bed.
- Several cycles of the circulation may be needed to build large clusters.
- As bed height is increased, the large circulation zone becomes more dominant and the possible residence time of a particle cluster in the bed becomes extended.
- Baffles can inhibit cluster formation as these clusters appear to be weakly bound together

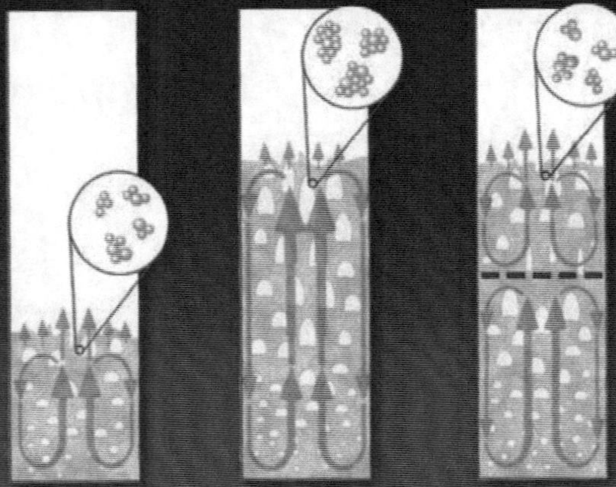

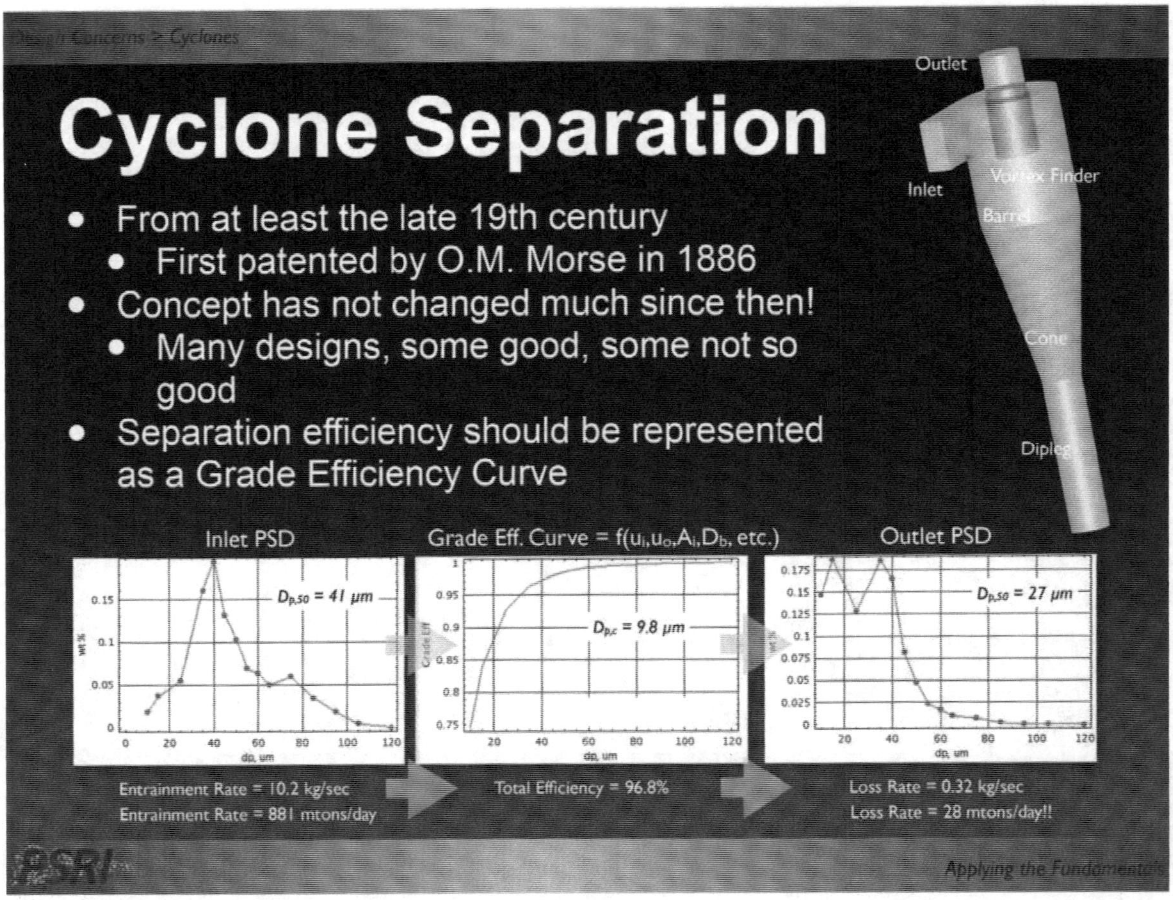
Design Concerns > Cyclones
Cyclone Separation
From at least the late 19th century
First patented by O.M. Morse in 1886
Concept has not changed much since then!
Many designs, some good, some not so good
Separation efficiency should be represented as a Grade Efficiency Curve
Outlet
Inlet
Vortex Finder
Barrel
Cone
Dipleg
Inlet PSD
Grade Eff. Curve = f(ui,uo,Ai,Db, etc.)
Outlet PSD
Dp,50 = 41 μm
Dp,c = 9.8 μm
Dp,50 = 27 μm
Entrainment Rate = 10.2 kg/sec
Entrainment Rate = 881 mtons/day
Total Efficiency = 96.8%
Loss Rate = 0.32 kg/sec
Loss Rate = 28 mtons/day!!
PSRI
Applying the Fundamentals

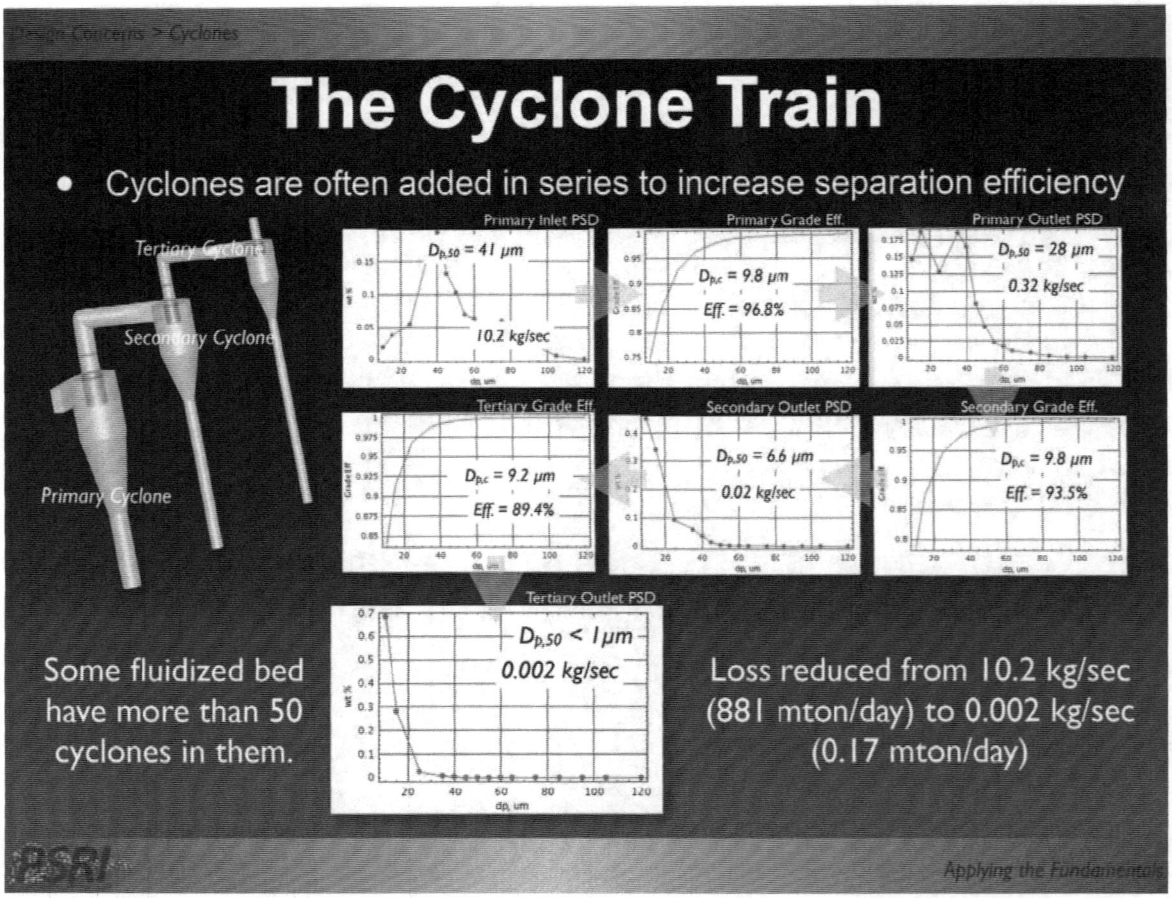
Design Concerns > Cyclones
The Cyclone Train
Cyclones are often added in series to increase separation efficiency
Tertiary Cyclone
Secondary Cyclone
Primary Cyclone
Primary Inlet PSD
Dp,50 = 41 μm
10.2 kg/sec
Primary Grade Eff.
Dp,c = 9.8 μm
Eff. = 96.8%
Primary Outlet PSD
Dp,50 = 28 μm
0.32 kg/sec
Tertiary Grade Eff.
Dp,c = 9.2 μm
Eff. = 89.4%
Secondary Outlet PSD
Dp,50 = 6.6 μm
0.02 kg/sec
Secondary Grade Eff.
Dp,c = 9.8 μm
Eff. = 93.5%
Tertiary Outlet PSD
Dp,50 < 1μm
0.002 kg/sec
Some fluidized bed have more than 50 cyclones in them.
Loss reduced from 10.2 kg/sec (881 mton/day) to 0.002 kg/sec (0.17 mton/day)
PSRI
Applying the Fundamentals

Particle Hydrodynamics in a Secondary Cyclone

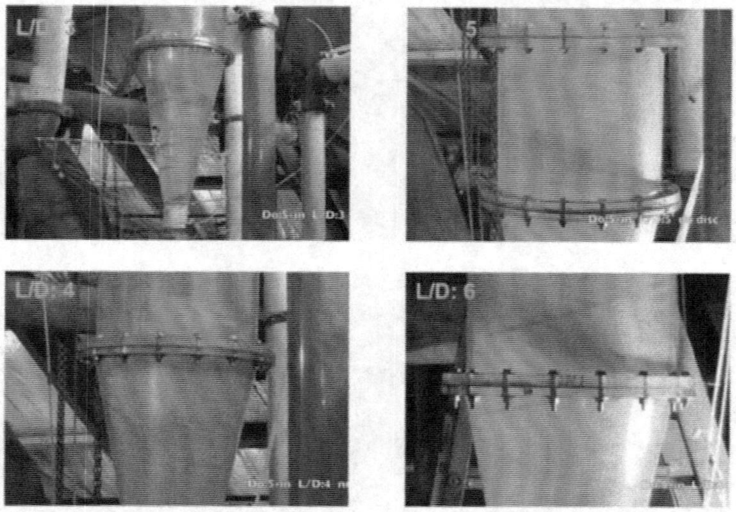

17-in (43-cm) barrel diameter cyclone with 65 ft/sec inlet velocity and 10 grains/ft³ (23 g/m³) loading using 27 micron FCC catalyst

Imaging the Inner Vortex

17-in (43-cm) dia. cyclone with 65 ft/sec (20 m/sec) inlet gas velocity and 10 gr/ft3 loading

L/D = 5

27 microns FCC catalyst

- The vortex stabilizer is preventing this inner vortex from precessing on the cone itself

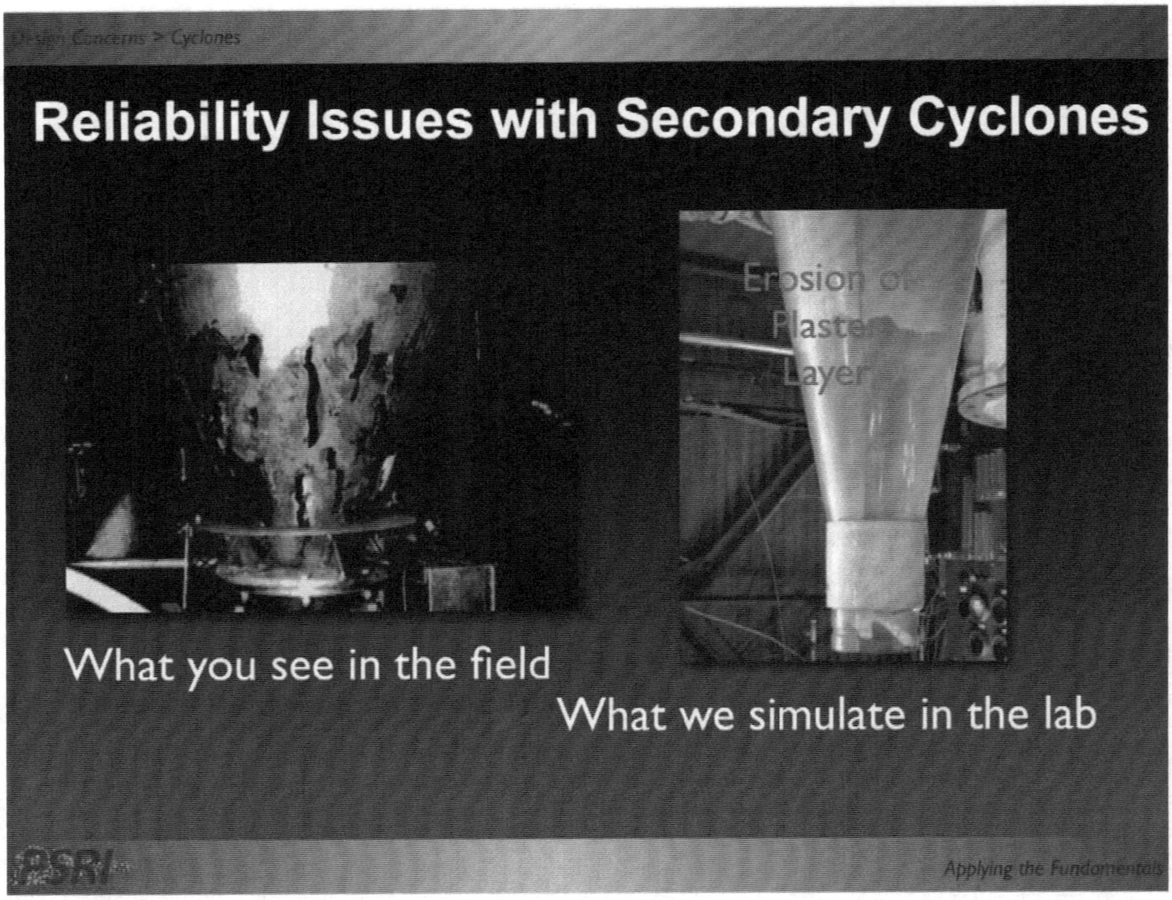
Design Concerns > Cyclones
Reliability Issues with Secondary Cyclones
What you see in the field
What we simulate in the lab
PSRI
Applying the Fundamentals

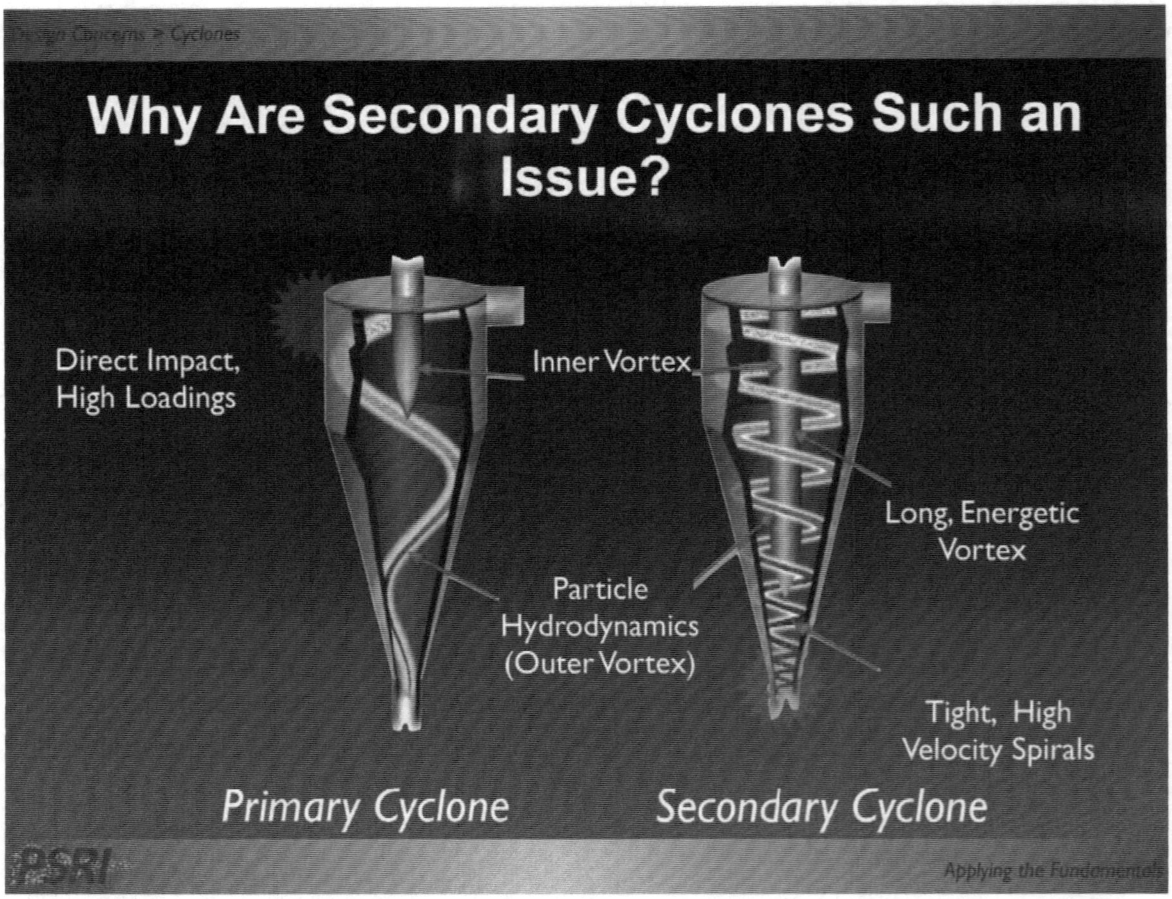
Design Concerns > Cyclones
Why Are Secondary Cyclones Such an Issue?
Direct Impact, High Loadings
Inner Vortex
Long, Energetic Vortex
Particle Hydrodynamics (Outer Vortex)
Tight, High Velocity Spirals
Primary Cyclone
Secondary Cyclone
PSRI
Applying the Fundamentals

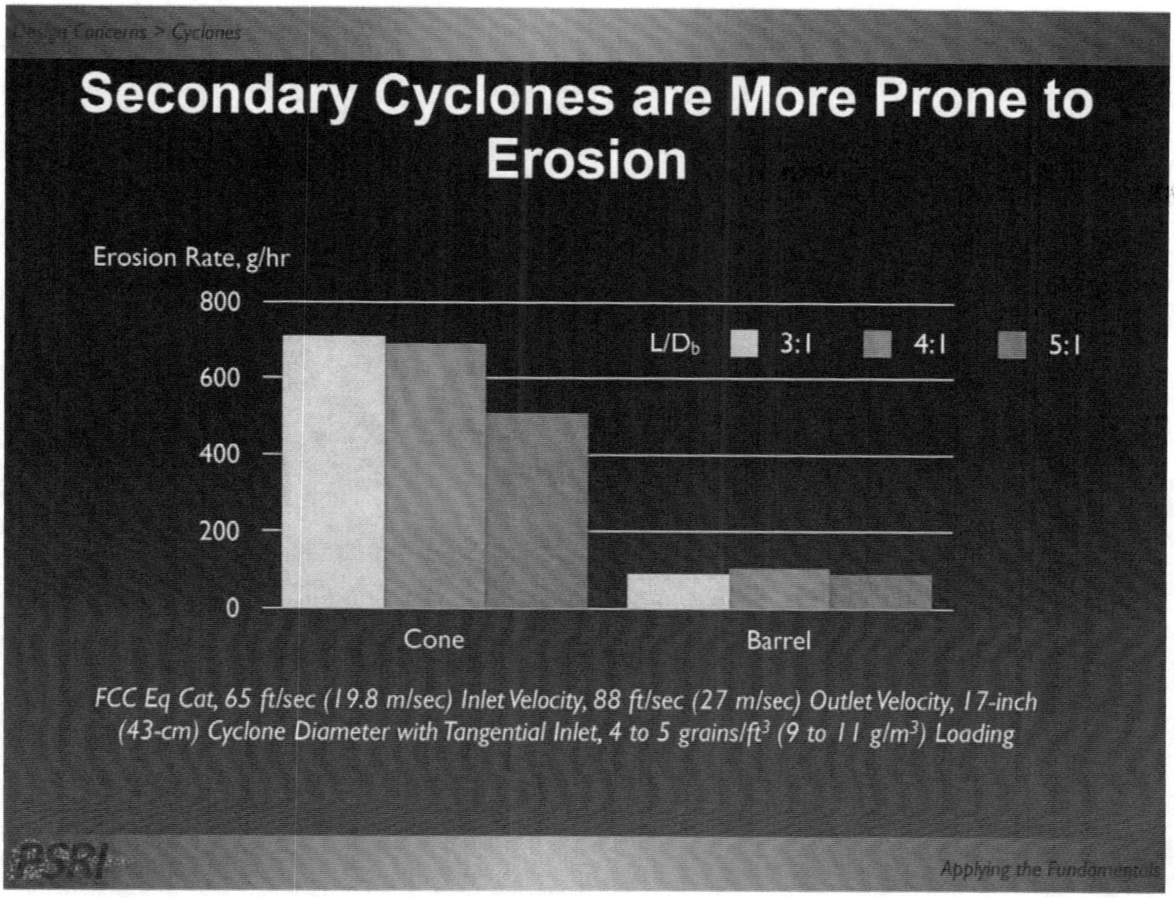
Design Concerns > Cyclones
Secondary Cyclones are More Prone to Erosion
Erosion Rate, g/hr
800
600
400
200
0
L/Db 3:1 4:1 5:1
Cone
Barrel
FCC Eq Cat, 65 ft/sec (19.8 m/sec) Inlet Velocity, 88 ft/sec (27 m/sec) Outlet Velocity, 17-inch (43-cm) Cyclone Diameter with Tangential Inlet, 4 to 5 grains/ft3 (9 to 11 g/m3) Loading
PSRI
Applying the Fundamentals

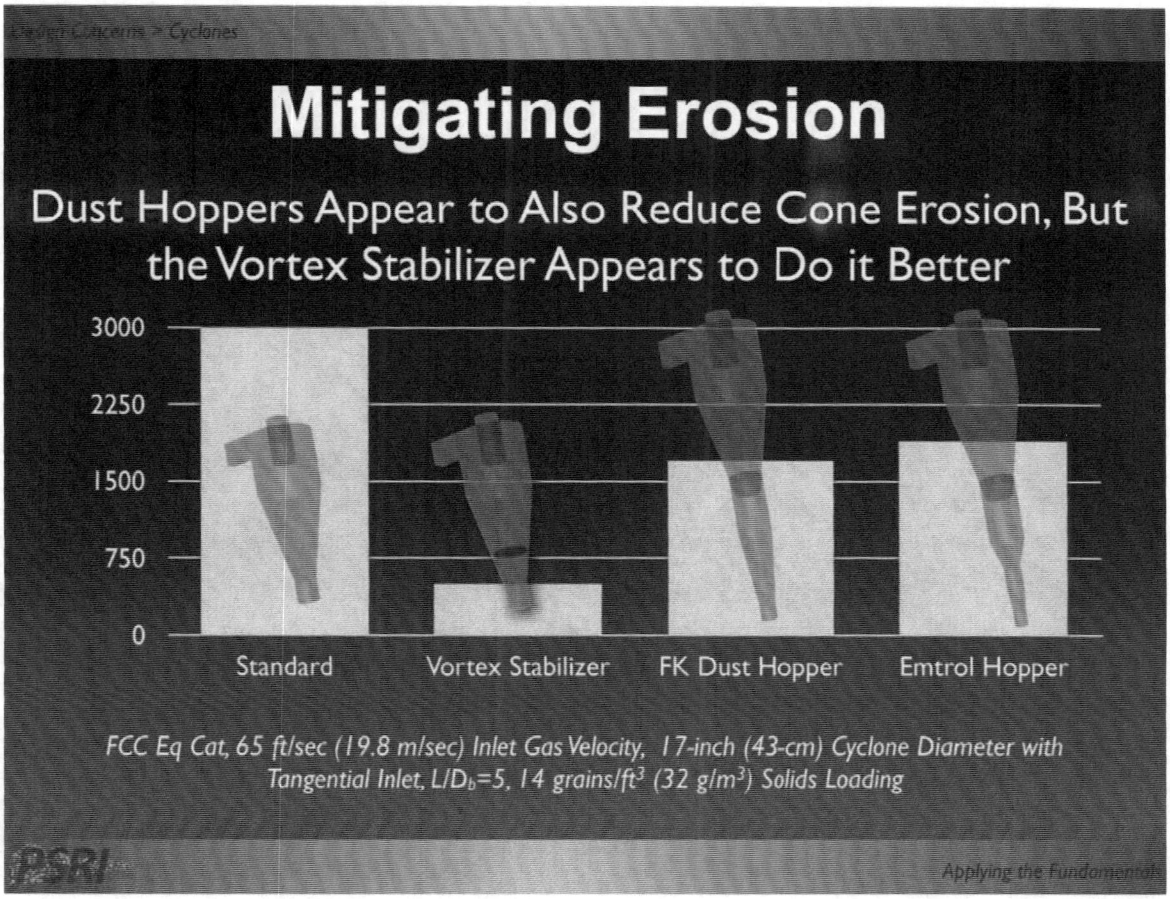
Design Concerns > Cyclones
Mitigating Erosion
Dust Hoppers Appear to Also Reduce Cone Erosion, But the Vortex Stabilizer Appears to Do it Better
3000
2250
1500
750
0
Standard
Vortex Stabilizer
FK Dust Hopper
Emtrol Hopper
FCC Eq Cat, 65 ft/sec (19.8 m/sec) Inlet Gas Velocity, 17-inch (43-cm) Cyclone Diameter with Tangential Inlet, L/Db=5, 14 grains/ft3 (32 g/m3) Solids Loading
PSRI
Applying the Fundamentals

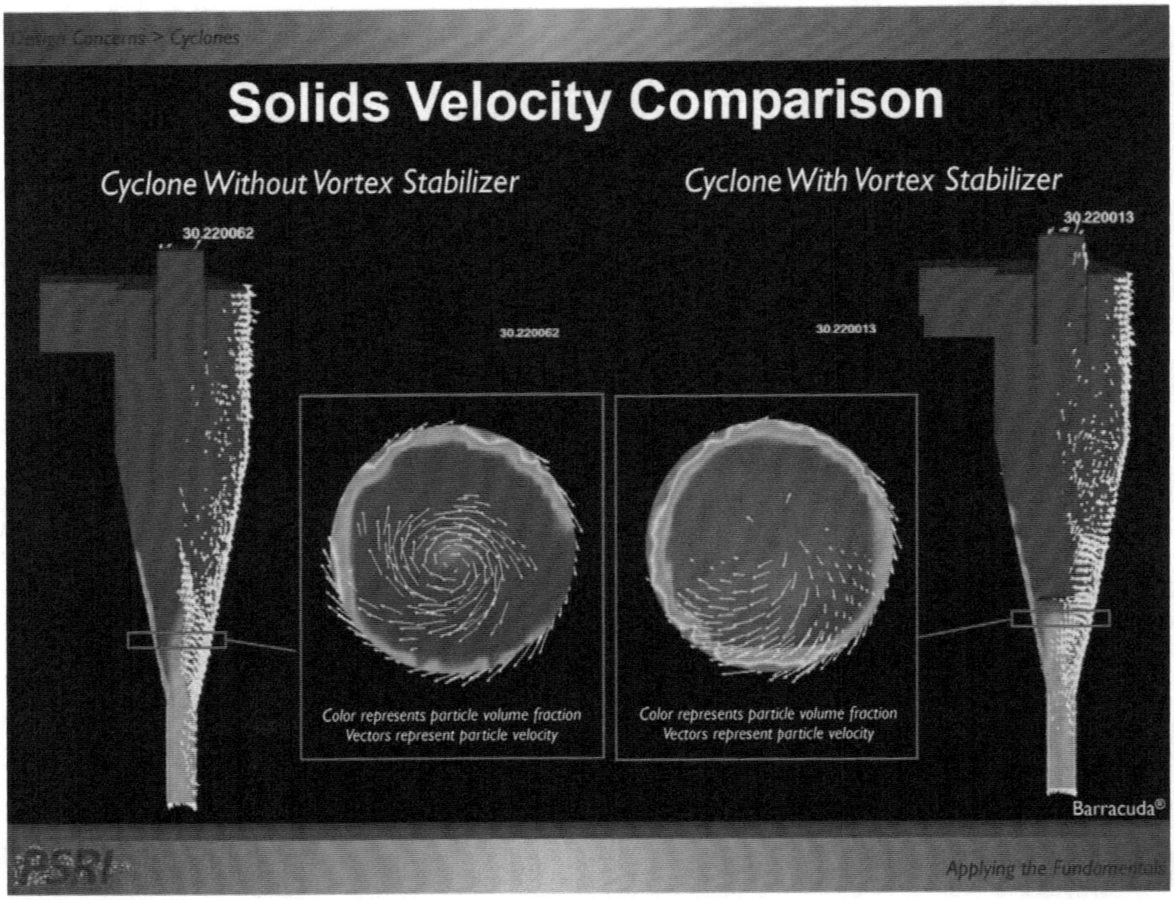
Design Concerns > Cyclones
Solids Velocity Comparison
Cyclone Without Vortex Stabilizer
Cyclone With Vortex Stabilizer
30.220062
30.220013
Color represents particle volume fraction
Vectors represent particle velocity
Color represents particle volume fraction
Vectors represent particle velocity
Barracuda®
PSRI
Applying the Fundamentals

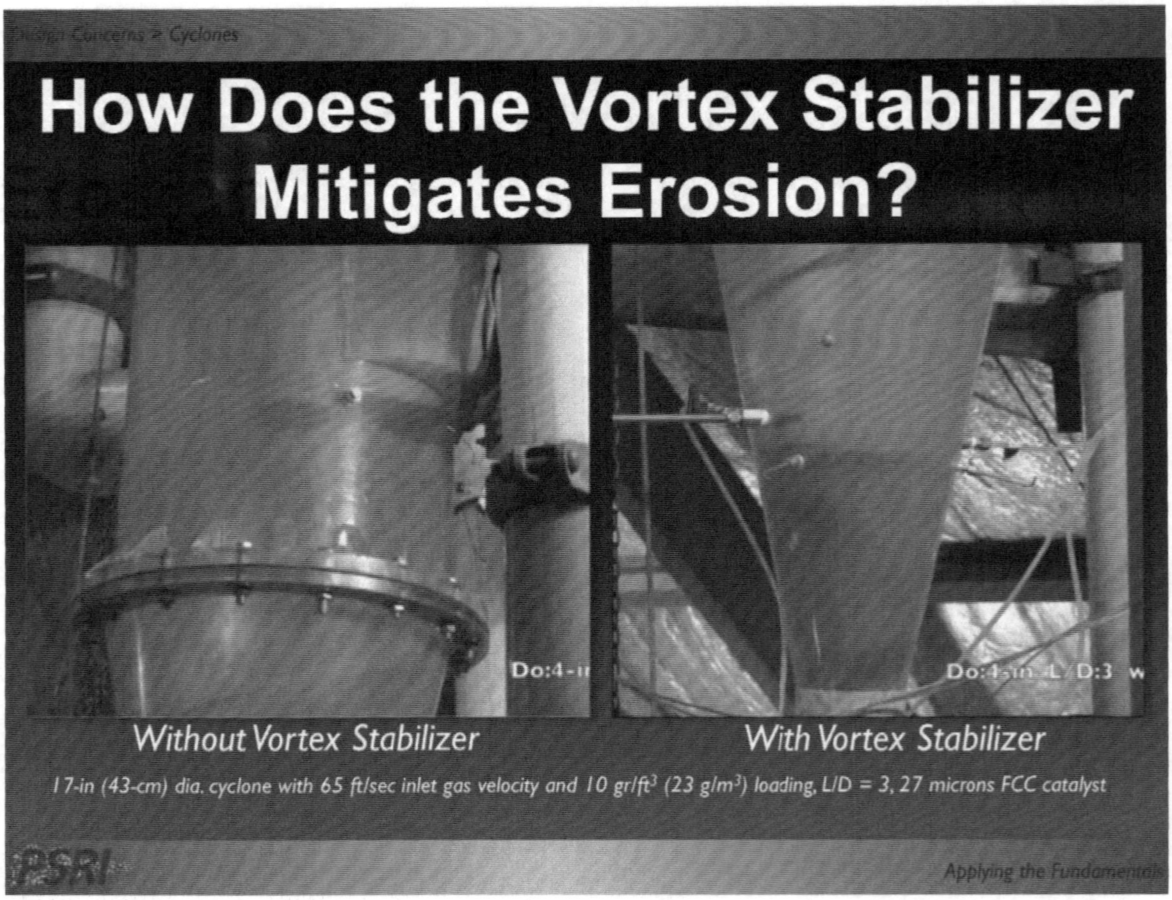
Design Concerns > Cyclones
How Does the Vortex Stabilizer Mitigates Erosion?
Without Vortex Stabilizer
With Vortex Stabilizer
17-in (43-cm) dia. cyclone with 65 ft/sec inlet gas velocity and 10 gr/ft^3 (23 g/m^3) loading, L/D = 3, 27 microns FCC catalyst
PSRI
Applying the Fundamentals

Summary

- Fluidized beds offer excellent heat transfer and the ability for continuous addition and withdrawal of solids
- Scale-up of fluidized beds is best done by understanding the underlying hydrodynamics including
  - Bed density
  - Bubble hydrodynamics (gas bypassing)
  - Particle clustering
  - Solids mixing
- Secondary cyclones can be a key reliability issue
  - Erosion in cone region
  - Can be mitigated with a vortex stabilizer

Ultrasonics / High Shear in Polymerization

Khungar, S., Middough

Originally presented on: 6/13/2013 15:30:00 - 16:00:00

Use of ultrasonics and high shear for polymer breakdown (degradation) is widely published. However, publications of their use during polymerization are not of consistent quality. The presenter's work on mass (bulk), solution, and micro-suspension and emulsion polymerization, using ultrasonics / high shear will be reported and compared with the published work. Scientific explanations will be given for the differences. The presenter has successfully developed processes which have saved millions of dollars. Examples will be given of lessons learned, including the following DOs and DON'Ts:

DOs:

- 1.Make good, accurate, scientific observations and repeat your work.
- 2.Find explanation for observed differences from your hypothesis, and change your hypothesis if needed, in view of the above scientific observations.
- 3.Make final 10-25% payment contingent on equipment performance validation.
- 4.Note surprising observations, find explanation and think of exploiting these.
- 5.Provide for contingencies, e.g., 10-50% extra horsepower for pumps, agitators, mixers, for future higher throughputs or higher viscosities, and consider the use of variable frequency drives (VFDs).
- 6.Work with willing equipment vendors to resolve in constructive ways, with creative solutions if the equipment underperforms.
- 7.Use established scale up parameters, such as constant tip speed (other choices are shear number or horsepower per unit volume).
- 8.Be open minded and receptive to input from others.

DON'Ts:

-
- 1.Trust all claims of equipment vendors without doing performance tests.
- 2.Trust all published work without duplicating or analyzing it carefully (DO challenge it where it seems wrong).
- 3.Believe in hearsay or opinions without corroborating from reliable sources.
- 4.Despair if you get negative/unexpected results (DO look for explanations).
- 5.Generalize, or get emotionally attached to a particular type of equipment for all applications, just because it worked wonderfully for one application. It may not work for the new application, or there may be a lower cost/ better alternative.

The Dos and Don'ts of Scaleup

Tatterson, G., North Carolina A&T State University
Originally presented on: 6/13/2013 16:00:00 - 16:30:00

The softer side of scaleup is reviewed in Dos and Don’ts of Scaleup. The important starting point, the importance of recognizing scale effects, different attitudes, approaches, methods and heuristics are presented. The scaleup ratio, geometric similarity, effective geometries, use of pilot plants, analogies, retrofits and myths are specifically examined.

AUTHOR INDEX

AIChE
120 Wall Street, FL 23
New York, NY 10005-4020

ISBN 978-1-5108-1604-6

CURRAN ASSOCIATES INC.
proceedings.com